ASE Test Preparation

Medium-Heavy Truck Certification Series

Gasoline Engines (T1)
5th Edition

DELMAR
CENGAGE Learning™

Australia • Brazil • Japan • Korea • Mexico • Singapore • Spain • United Kingdom • United States

ASE Test Preparation: Medium-Heavy Truck Certification Series, Gasoline Engines (T1), 5th Edition

Vice President, Technology and Trades Professional Business Unit: Gregory L. Clayton

Director, Professional Transportation Industry Training Solutions: Kristen L. Davis

Editorial Assistant: Danielle Filippone

Director of Marketing: Beth A. Lutz

Marketing Manager: Jennifer Barbic

Senior Production Director: Wendy Troeger

Production Manager: Sherondra Thedford

Content Project Management: PreMediaGlobal

Senior Art Director: Benjamin Gleeksman

Section Opener Image: Image Copyright Goran Bogicevic, 2012. Used under license from Shutterstock.com

ISBN-13: 978-1-111-12897-5

ISBN-10: 1-111-12897-9

Delmar Cengage Learning
5 Maxwell Drive
Clifton Park, NY 12065-2919
USA

Cengage Learning is a leading provider of customized learning solutions with office locations around the globe, including Singapore, the United Kingdom, Australia, Mexico, Brazil, and Japan. Locate your local office at: **international.cengage.com/region**.

Cengage Learning products are represented in Canada by Nelson Education, Ltd.

For more information on transportation titles available from Delmar, Cengage Learning, please visit our website at **www.trainingbay.cengage.com**.

For your lifelong learning solutions, visit **delmar.cengage.com**.

Visit our corporate website at **www.cengage.com**.

Printed in the United States of America
1 2 3 4 5 6 7 16 15 14 13 12 11

Table of Contents

Preface v

SECTION 1 **The History and Purpose of ASE** 1

SECTION 2 **Overview and Introduction** **2**

Exam Administration 2

Understanding Test Question Basics 3

Test-Taking Strategies 3

Preparing for the Exam 4

What to Expect During the Exam 6

Testing Time. 7

Understanding How Your Exam Is Scored 8

SECTION 3 **Types of Questions on an ASE Exam** **9**

Multiple-Choice Questions 9

Technician A, Technician B Questions 10

EXCEPT Questions 11

LEAST LIKELY Questions. 12

Summary 12

SECTION 4 **Task List Overview** **13**

Introduction 13

SECTION 5 **Sample Preparation Exams** **61**

Introduction 61

Preparation Exam 1 61

Preparation Exam 2 69

Preparation Exam 3 78

Preparation Exam 4 87

Preparation Exam 5 97

Preparation Exam 6 107

SECTION 6   Answer Keys and Explanations 118

Introduction 118

Preparation Exam 1—Answer Key 118

Preparation Exam 1—Explanations 119

Preparation Exam 2—Answer Key 136

Preparation Exam 2—Explanations 136

Preparation Exam 3—Answer Key 154

Preparation Exam 3—Explanations 154

Preparation Exam 4—Answer Key 172

Preparation Exam 4—Explanations 172

Preparation Exam 5—Answer Key 191

Preparation Exam 5—Explanations 191

Preparation Exam 6—Answer Key 211

Preparation Exam 6—Explanations 211

SECTION 7   Appendices 230

Preparation Exam Answer Sheet Forms 230

Glossary 236

Preface

Delmar, a part of Cengage Learning, is very pleased that you have chosen to use our ASE Test Preparation Guide to help prepare yourself for the Gasoline Engines (T1) ASE certification examination. This guide is designed to help prepare you for your actual exam by providing you with an overview and introduction of the testing process, introducing you to the task list for the Gasoline Engines (T1) certification exam, giving you an understanding of what knowledge and skills you are expected to have in order to successfully perform the duties associated with each task area, and providing you with several preparation exams designed to emulate the live exam content in hopes of assessing your overall exam readiness.

If you have a basic working knowledge of the discipline you are testing for, you will find this book is an excellent guide, helping you understand the "must know" items needed to successfully pass the ASE certification exam. This manual is not a textbook. Its objective is to prepare the individual who has the existing requisite experience and knowledge to attempt the challenge of the ASE certification process. This guide cannot replace the hands-on experience and theoretical knowledge required by ASE to master the vehicle repair technology associated with this exam. If you are unable to understand more than a few of the preparation questions and their corresponding explanations in this book, it could be that you require either more shop-floor experience or further study.

This book begins by providing an overview of, and introduction to, the testing process. This section outlines what we recommend you do to prepare, what to expect on the actual test day, and overall methodologies for your success. This section is followed by a detailed overview of the ASE task list to include explanations of the knowledge and skills you must possess to successfully answer questions related to each particular task. After the task list, we provide six sample preparation exams for you to use as a means of evaluating areas of understanding, as well as areas requiring improvement in order to successfully pass the ASE exam. Delmar is the first and only test preparation organization to provide so many unique preparation exams. We enhanced our guides to include this support as a means of providing you with the best preparation product available. Section 6 of this guide includes the answer keys for each preparation exam, along with the answer explanations for each question. Each answer explanation also contains a reference back to the related task or tasks that it assesses. This will provide you with a quick and easy method for referring back to the task list whenever needed. The last section of this book contains blank answer sheet forms you can use as you attempt each preparation exam, along with a glossary of terms.

OUR COMMITMENT TO EXCELLENCE

Thank you for choosing Delmar, Cengage Learning for your ASE test preparation needs. All of the writers, editors, and Delmar staff have worked very hard to make this test preparation guide second to none. We feel confident that you will find this guide easy to use and extremely beneficial as you prepare for your actual ASE exam.

Delmar, Cengage Learning has sought out the best subject-matter experts in the country to help with the development of *ASE Test Preparation: Medium-Heavy Truck Certification Series, Gasoline Engines (T1), 5th Edition*. Preparation questions are authored and then reviewed by a group of certified, subject-matter experts to ensure the highest level of quality and validity to our product.

If you have any questions concerning this guide or any guide in this series, please visit us on the web at **http://www.trainingbay.cengage.com**.

For web-based online test preparation for ASE certifications, please visit us on the web at **http://www.techniciantestprep.com/** to learn more.

ABOUT THE SERIES ADVISOR

Brian (BJ) Crowley has experienced several different aspects of the diesel industry over the past ten years. Now a diesel technician in the oil and gas industry, BJ owned and operated a diesel repair shop where he repaired heavy, medium, and light trucks—in addition to agricultural and construction equipment. He earned an Associate's degree in diesel technology from Elizabethtown Community and Technical College and is an ASE Master certified medium/heavy truck technician.

The History and Purpose of ASE

ASE began as the National Institute for Automotive Service Excellence (NIASE). It was founded as a non-profit, independent entity in 1972 by a group of industry leaders with the single goal of providing a means for consumers to distinguish between incompetent and competent technicians. It accomplishes this goal through the testing and certification of repair and service professionals. Though it is still known as the National Institute for Automotive Service Excellence, it is now called "ASE" for short.

Today, ASE offers more than 40 certification exams in automotive, medium/heavy duty trucks, collision repair and refinish, school bus, transit bus, parts specialist, automobile service consultant, and other industry-related areas. At this time, there are more than 385,000 professionals nationwide with current ASE certifications. These professionals are employed by new car and truck dealerships, independent repair facilities, fleets, service stations, franchised service facilities, and more.

ASE's certification exams are industry-driven and cover practically every on-highway vehicle service segment. The exams are designed to stress the knowledge of job-related skills. Certification consists of passing at least one exam and documenting two years of relevant work experience. To maintain certification, those with ASE credentials must be re-tested every five years.

While ASE certifications are a targeted means of acknowledging the skills and abilities of an individual technician, ASE also has a program designed to provide recognition for highly qualified repair, support, and parts businesses. The Blue Seal of Excellence Recognition Program allows businesses to showcase their technicians and their commitment to excellence. One of the requirements of becoming Blue Seal recognized is that the facility must have a minimum of 75 percent of their technicians ASE certified. Additional criteria apply, and program details can be found on the ASE website.

ASE recognized that educational programs serving the service and repair industry also needed a way to be recognized as having the faculty, facilities, and equipment to provide a quality education to students wanting to become service professionals. Through the combined efforts of ASE, industry, and education leaders, the non-profit organization entitled the National Automotive Technicians Education Foundation (NATEF) was created in 1983 to evaluate and recognize academic programs. Today more than 2,000 educational programs are NATEF certified.

For additional information about ASE, NATEF, or any of their programs, the following contact information can be used:

National Institute for Automotive Service Excellence (ASE)

101 Blue Seal Drive S.E.

Suite 101

Leesburg, VA 20175

Telephone: 703-669-6600

Fax: 703-669-6123

Website: **www.ase.com**

Participating in the National Institute for Automotive Service Excellence (ASE) voluntary certification program provides you with the opportunity to demonstrate you are a qualified and skilled professional technician who has the "know-how" required to successfully work on today's modern vehicles.

EXAM ADMINISTRATION

Through 2011, there are two methods available to you when taking an ASE certification exam:

- Paper and pencil
- Computer Based Testing (CBT)

> *Note:* Beginning 2012, ASE will no longer offer paper and pencil certification exams. They will offer and support CBT testing exclusively.

Paper and Pencil Exams

ASE paper and pencil exams are administered twice annually, once in the spring and once again in the fall. The paper and pencil exams are administered at over 750 exam sites in local communities across the nation.

Each test participant is given a booklet containing questions with charts and diagrams where required. All instructions are printed on the exam materials and should be followed carefully. You can mark in this exam booklet but no information entered in the booklet is scored. You will record your answers using a separate answer sheet. You will need to mark your answers, using a #2 pencil only. Upon completion of your exam, the answer sheets are electronically scanned and the answers are tabulated.

> *Note:* Paper and pencil exams will no longer be offered by ASE after 2011. ASE will be converting to a completely exclusive CBT testing methodology at that time.

CBT Exams

ASE also provides CBT exams, which are administered twice annually, once in the winter and once again in the summer. The CBT exams are administered at test centers across the nation. The exam content is the same for both the paper and pencil and CBT testing methods.

If you are considering the CBT exams, it is recommended that you go to the ASE website at ***http://www.ase.com*** and review the conditions and requirements for this type of exam. There is also an exam demonstration page that allows you to personally experience how this type of exam operates before you register.

Effective 2012, ASE will only offer CBT testing. At that time, CBT exams will be available four times annually, for two-month windows, with a month of no testing in between each testing window:

- January/February – Winter CBT testing window
- April/May – Spring CBT testing window
- July/August – Summer CBT testing window
- October/November – Fall CBT testing window

Please note, testing windows and timing may change. It is recommended you go to the ASE website at *http://www.ase.com* and review the latest testing schedules.

UNDERSTANDING TEST QUESTION BASICS

ASE exam questions are written by service industry experts. Each question on an exam is created during an ASE-hosted "item-writing" workshop. During these workshops, expert service representatives from manufacturers (domestic and import), aftermarket parts and equipment manufacturers, working technicians, and technical educators gather to share ideas and convert them into actual exam questions. Each exam question written by these experts must then survive review by all members of the group. The questions are designed to address the practical application of repair and diagnosis knowledge and skills practiced by technicians in their day-to-day work.

After the item-writing workshop, all questions are pre-tested and quality-checked on a national sample of technicians. Those questions that meet ASE standards of quality and accuracy are included in the scored sections of the exams; the "rejects" are sent back to the drawing board or discarded altogether.

Depending on the topic of the certification exam, you will be asked between 40 and 80 multiple-choice questions. You can determine the approximated number of questions you can expect to be asked during the Gasoline Engines (T1) certification exam by reviewing the task list in Section 4 of this book. The five-year recertification exam will cover this same content; however, the number of questions for each content area of the recertification exam will be reduced by approximately one-half.

> *Note:* Exams may contain questions that are included for statistical research purposes only. Your answers to these questions will not affect your score, but since you do not know which ones they are, you should answer all questions in the exam.

Using multiple criteria, including cross-sections by age, race, and other background information, ASE is able to guarantee that exam questions do not include bias for or against any particular group. A question that shows bias toward any particular group is discarded.

TEST-TAKING STRATEGIES

Before beginning your exam, quickly look over the exam to determine the total number of questions that you will need to answer. Having this knowledge will help you manage your time throughout the exam to ensure you have enough available to answer all of the questions presented. Read through each question completely before marking your answer. Answer the questions in the order they appear on the exam. Leave the questions blank that you are not sure of and move on to the next question. You can return to those unanswered questions after you have finished the others. These questions may actually be easier to answer at a later time once your mind has had additional time to consider them on a subconscious level. In addition, you might find information in other questions that will help you recall the answers to some of them.

Multiple-choice exams are sometimes challenging because there are often several choices that may seem possible, or partially correct, and therefore it may be difficult to decide on the most appropriate answer choice. The best strategy, in this case, is to first determine the correct answer before looking at the answer options. If you see the answer you decided on, you should still be careful to examine the other answer options to make sure that none seems more correct than yours. If you do not know or are not sure of the answer, read each option very carefully and try to eliminate those options that you know are incorrect. That way, you can often arrive at the correct choice through a process of elimination.

If you have gone through the entire exam, and you still do not know the answer to some of the questions, *then guess*. Yes, guess. You then have at least a 25 percent chance of being correct. While your score is based on the number of questions answered correctly, any question left blank, or unanswered, is automatically scored as incorrect.

There is a lot of "folk" wisdom on the subject of test taking that you may hear about as you prepare for your ASE exam. For example, there are those who would advise you to avoid response options that use certain words such as *all, none, always, never, must,* and *only,* to name a few. This, they claim, is because nothing in life is exclusive. They would advise you to choose response options that use words that allow for some exception, such as *sometimes, frequently, rarely, often, usually, seldom,* and *normally.* They would also advise you to avoid the first and last option (A or D) because exam writers, they feel, are more comfortable if they put the correct answer in the middle (B or C) of the choices. Another recommendation often offered is to select the option that is either shorter or longer than the other three choices because it is more likely to be correct. Some would advise you to never change an answer since your first intuition is usually correct. Another area of "folk" wisdom focuses specifically on any repetitive patterns created by your question responses (e.g., A, B, C, A, B, C, A, B, C).

Many individuals may tell that there are actual grains of truth in this "folk" wisdom, and whereas with some exams, this may prove true, it is not relevant in regard to the ASE certification exams. ASE validates all exam questions and test forms through a national sample of technicians, and only those questions and test forms that meet ASE standards of quality and accuracy are included in the scored sections of the exams. Any biased questions or patterns are discarded altogether, and therefore, it is highly unlikely you will experience any of this "folk" wisdom on an actual ASE exam.

PREPARING FOR THE EXAM

Delmar, Cengage Learning wants to make sure we are providing you with the most thorough preparation guide possible. To demonstrate this, we have included hundreds of preparation questions in this guide. These questions are designed to provide as many opportunities as possible to prepare you to successfully pass your ASE exam. The preparation approach we recommend and outline in this book is designed to help you build confidence in demonstrating what task area content you already know well while also outlining what areas you should review in more detail prior to the actual exam.

We recommend that your first step in the preparation process should be to thoroughly review Section 3 of this book. This section contains a description and explanation of the type of questions you'll find on an ASE exam.

Once you understand how the questions will be presented, we then recommend that you thoroughly review Section 4 of this book. This section contains information that will help you establish an understanding of what the exam will be evaluating, and specifically, how many questions to expect in each specific task area.

As your third preparatory step, we recommend you complete your first preparation exam, located in Section 5 of this book. Answer one question at a time. After you answer each question, review the answer and question explanation information, located in Section 6. This section will provide you with instant response feedback, allowing you to gauge your progress, one question at a time, throughout this first preparation exam. If after reading the question explanation you don't feel you understand the reasoning for the correct answer, go back and review the task list overview (Section 4) for the task that is related to that question. Included with each question explanation is a clear identifier of the task area that is being assessed (e.g., Task A.1). If at that point you still don't feel you have a solid understanding of the material, identify a good source of information on the topic, such as an educational course, textbook or other related source of topical learning, and do some additional studying.

After you have completed your first preparation exam and have reviewed your answers, you are ready to complete your next preparation exam. A total of six practice exams are available in Section 5 of this book. For your second preparation exam, we recommend that you answer the questions as if you were taking the actual exam. Do not use any reference material or allow any interruptions in order to get a feel for how you will do on the actual exam. Once you have answered all of the questions, grade your results using the answer key in Section 6. For every question that you gave an incorrect answer to, study the explanations to the answers and/or the overview of the related task areas. Try to determine the root cause for missing the question. The easiest thing to correct is learning the correct technical content. The hardest things to correct are behaviors that lead you to an incorrect conclusion. If you knew the information but still got the question incorrect, there is likely a test-taking behavior that will need to be corrected. An example of this would be reading too quickly and skipping over words that affect your reasoning. If you can identify what you did that caused you to answer the question incorrectly, you can eliminate that cause and improve your score.

Here are some basic guidelines to follow while preparing for the exam:

- Focus your studies on those areas you are weak in.
- Be honest with yourself when determining if you understand something.
- Study often but for short periods of time.
- Remove yourself from all distractions when studying.
- Keep in mind that the goal of studying is not just to pass the exam; the real goal is to learn.
- Prepare physically by getting a good night's rest before the exam, and eat meals that provide energy but do not cause discomfort.
- Arrive early to the exam site to avoid long waits as test candidates check in.
- Use all of the time available for your exams. If you finish early, spend the remaining time reviewing your answers.
- Do not leave any questions unanswered. If absolutely necessary, guess. All unanswered questions are automatically scored as incorrect.

Here are some items you will need to bring with you to the exam site:

- A valid government or school-issued photo ID
- Your test center admissions ticket
- Three or four sharpened #2 pencils and an eraser
- A watch (not all test sites have clocks)

> *Note:* Books, calculators, and other reference materials are not allowed in the exam room. The exceptions to this list are English-Foreign dictionaries, or glossaries. All items will be inspected before and after testing.

WHAT TO EXPECT DURING THE EXAM

Paper and Pencil Exams

When taking a paper and pencil exam, you will be placing your answers on a sheet that requires you to blacken (bubble) in your answer choice.

Be careful that only your answers are visible on the answer sheet. Stray pencil marks or incomplete erasures may be picked up as an answer by the electronic reader and result in a question being scored incorrectly.

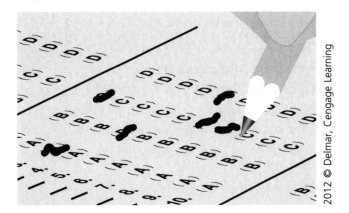

Studies have shown that one of the biggest challenges an adult faces when taking a test that uses a bubble-style answer sheet is to place their answers in the correct location. To avoid problems in this area, be extra mindful of how and where you mark your answers. For example, when answering question 21, blacken the correct, corresponding bubble on the answer sheet for question 21. Pay special attention to this process when you decide to skip a question to come back to later. In this situation, many people forget to also leave the corresponding line on the bubble answer sheet blank as well. They inadvertently place their answer for the next question on the answer bubble sheet line that should have been left as a blank placeholder for the unanswered, skipped question. Providing a correct question response on the incorrect bubble answer sheet line will likely result in that question being marked wrong. Remember, the answer sheet for the paper and pencil exam is machine scored, and the machine can only "read" what you have blackened or bubbled in.

If you finish answering all of the questions on an exam and have time remaining, go back and review the answers for those questions that you were not sure of. You can often catch careless errors by using the remaining time to review your answers. Carefully check your answer sheet for blank answers or missing information.

At practically every exam, some technicians will finish ahead of time and turn their papers in long before the final call. Since some technicians may be doing a recertification test and others may be taking fewer exams than you, do not let this distract or intimidate you.

It is not wise to use less than the total amount of time that you are allotted for an exam. If there are any doubts, take the time for review. Any product can usually be made better with some additional effort. An exam is no exception. It is not necessary to turn in your exam paper until you are told to do so.

CBT Exams

When taking a CBT exam, as soon as you are seated in the testing center, you will be given a brief tutorial to acquaint you with the computer-delivered test, prior to taking your certification exam(s). Unlike paper and pencil testing, when taking a CBT exam, you will not have to worry about stray pencil marks or ensuring that your answers are marked on the correct and corresponding answer bubble sheet line. The CBT exams allow you to select only one answer per question. You can also change your answers as many times as you like. When you select a second answer choice, the CBT will automatically unselect your first answer choice. If you want to skip a question to return to later, you can utilize the "flag" feature, which will allow you to quickly identify and review questions whenever you are ready. Prior to completing your exam, you will also be provided with an opportunity to review your answers and address any unanswered questions.

TESTING TIME

Paper and Pencil Exams

Each ASE paper and pencil exam session is four hours. You may register for and take anywhere from one to a maximum of four exams during any one exam session. It is recommended, however, that you do not register for any combination of exams that would result in you having to answer any more than 225 questions during any single exam session. As a worst-case scenario, this will allow you only slightly more than one minute to answer each question.

CBT Exams

Unlike the ASE paper and pencil exams, each individual ASE CBT exam has a fixed time limit. Individual exam times will vary based upon exam area, and will range anywhere from a half hour to two hours. You will also be given an additional 30 minutes beyond what is allotted to complete your exams to ensure you have adequate time to perform all necessary check-in procedures, complete a brief CBT tutorial, and potentially complete a post-test survey.

Similar to the paper and pencil exams, you can register for and take multiple CBT exams during one testing appointment. The maximum time allotment for a CBT appointment is four and a half hours. If you happen to register for so many exams that you will require more time than this, your exams will be scheduled into multiple appointments. This could mean that you have testing on both the morning and afternoon of the same day, or they could be scheduled on different days, depending on your personal preference and the test center's schedule.

It is important to understand that if you arrive late for your CBT test appointment, you will not be able to make up any missed time. You will only have the scheduled amount of time remaining in your appointment to complete your exam(s).

Also, while most people finish their CBT exams within the time allowed, others might feel rushed or not be able to finish the test, due to the implied stress of a specific, individual time limit allotment. Before you register for the CBT exams, you should review the number of exam questions that will be asked along with the amount of time allotted for that exam to determine whether you feel comfortable with the designated time limitation or not.

Summary

Regardless of whether you are taking a paper and pencil or CBT exam, as an overall time management recommendation, you should monitor your progress and set a time limit you will follow with regard to how much time you will spend on each individual exam question. This should be based on the total number of questions you will be answering.

Also, it is very important to note that if for any reason you wish to leave the testing room during an exam, you must first ask permission. If you happen to finish your exam(s) early and wish to leave the testing site before your designated session appointment is completed, you are permitted to do so only during specified dismissal periods.

UNDERSTANDING HOW YOUR EXAM IS SCORED

You can gain a better perspective about the ASE certification exams if you understand how they are scored. ASE exams are scored by an independent organization having no vested interest in ASE or in the automotive industry.

Each question carries the same weight as any other question. For example, if there are 50 questions, each is worth 2 percent of the total score. Your exam results can tell you

- Where your knowledge equals or exceeds that needed for competent performance, or
- Where you might need more preparation.

Your ASE exam score report is divided into content "task" areas; it will show the number of questions in each content area and how many of your answers were correct. These numbers provide information about your performance in each area of the exam. However, because there may be a different number of questions in each content area of the exam, a high percentage of correct answers in an area with few questions may not offset a low percentage in an area with many questions.

It should be noted that one does not "fail" an ASE exam. The technician who does not pass is simply told "More Preparation Needed." Though large differences in percentages may indicate problem areas, it is important to consider how many questions were asked in each area. Since each exam evaluates all phases of the work involved in a service specialty, you should be prepared in each area. A low score in one area could keep you from passing an entire exam.

There is no such thing as average. You cannot determine your overall exam score by adding the percentages given for each task area and dividing by the number of areas. It doesn't work that way because there generally are not the same number of questions in each task area. A task area with 20 questions, for example, counts more toward your total score than a task area with 10 questions.

Your exam report should give you a good picture of your results and a better understanding of your strengths and areas needing improvement for each task area.

If you fail to pass the exam, you may take it again at any time it is scheduled to be administered. You are the only one who will receive your exam score. Exam scores will not be given over the telephone by ASE nor will they be released to anyone without your written permission.

Types of Questions on an ASE Exam

Understanding not only what content areas will be assessed during your exam, but how you can expect exam questions to be presented will enable you to gain the confidence you need to successfully pass an ASE certification exam. The following examples will help you recognize the types of question styles used in ASE exams and assist you in avoiding common errors when answering them.

Most initial certification tests are made up of between 40 to 80 multiple-choice questions. The five-year recertification exams will cover the same content as the initial exam; however, the actual number of questions for each content area will be reduced by approximately one-half. Refer to Section 4 of this book for specific details regarding the number of questions to expect during the initial Gasoline Engines (T1) certification exam.

Multiple-choice questions are an efficient way to test knowledge. To correctly answer them, you must consider each answer choice as a possibility, and then choose the answer choice that *best* addresses the question. To do this, read each word of the question carefully. Do not assume you know what the question is asking until you have finished reading the entire question.

About 10 percent of the questions on an actual ASE exam will reference an illustration. These drawings contain the information needed to correctly answer the question. The illustration should be studied carefully before attempting to answer the question. When the illustration is showing a system in detail, look over the system and try to figure out how the system works before you look at the question and the possible answers. This approach will ensure that you do not answer the question based upon false assumptions or partial data, but instead have reviewed the entire scenario being presented.

MULTIPLE-CHOICE QUESTIONS

The most common type of question used on an ASE exam is direct completion, which is more commonly referred to as a multiple-choice style question. This type of question contains three "distracters" (incorrect answers) and one "key" (correct answer). When the questions are written, the point is to make the distracters plausible to draw an inexperienced technician to inadvertently select one of them. This type of question gives a clear indication of the technician's knowledge.

Examples of this type of question would appear as follows:

1. Which of the following would be used to measure crankshaft main bearing clearance on a gasoline engine?

   A. Dial indicator

   B. Plastigauge®

   C. Outside micrometer

   D. Dial caliper

Answer A is incorrect. A dial indicator is used to measure crankshaft end-play or runout. It would not be useful to measure crankshaft main bearing clearance.

Answer B is correct. Plastigauge is the most common method of measuring crankshaft main bearing clearance.

Answer C is incorrect. An outside micrometer is used to measure the outside dimension of a part; however, alone it cannot measure crankshaft main bearing clearance.

Answer D is incorrect. A dial caliper is used to measure outside depth or inside dimensions, but it would not be an effective tool to measure crankshaft main bearing clearance.

2. A gasoline engine has an accelerator pedal position (APP) diagnostic trouble code. The technician wiggles the wiring harness while observing the accelerator pedal position sensor voltage with the scan tool. The sensor voltage changes. Which of the following is the most likely cause of the diagnostic trouble code?

    A. Faulty scan tool
    B. Faulty ECM
    C. Faulty APP sensor wiring
    D. Faulty ECM power supply

Answer A is incorrect. If the voltage value changed while moving the wiring harness, there is no reason to believe the scan tool is faulty. If the scan tool would not communicate with the ECM, then there is a possibility that the scan tool is faulty.

Answer B is incorrect. If the voltage value changed while moving the wiring harness, there is no reason to believe the ECM is faulty. If the ECM did not send an APP voltage signal to the scan tool, then the ECM may be faulty.

Answer C is correct. If the voltage value changed while moving the wiring harness, the most likely cause is the wiring harness.

Answer D is incorrect. If all other items associated with the ECM are normal and the voltage value changed while moving the wiring harness, the most likely cause is the wiring harness.

TECHNICIAN A, TECHNICIAN B QUESTIONS

The type of question style that is most popularly associated with an ASE exam is the "Technician A says … Technician B says … Who is right?" type of question. In this type of question, you must identify the correct statement or statements. To answer this type of question correctly, you must carefully read each technician's statement and judge it on its own merit to determine if the statement is true.

Sometimes this type of question begins with a statement about some analysis or repair procedure. This is often referred to as the stem of the question and provides the setup or background information required to understand the conditions on which the question is based. This is followed by two statements about the cause of the concern, proper inspection, identification, or repair choices. You are asked whether the first statement, the second statement, both statements, or neither statement is correct. Analyzing this type of question is a little easier than the other types because there are only two ideas to consider, although there are still four choices for an answer.

Technician A, Technician B questions are really double true-or-false questions. The best way to analyze this type of question is to consider each technician's statement separately. Ask yourself, is A true or false? Is B true or false? Once you have completed this individual evaluation of each answer choice, you will have successfully determined the correct answer choice. An important point to remember is that an ASE Technician A, Technician B question will never have Technician A and B directly disagreeing with each other. That is why you must evaluate each statement independently.

An example of this type of question would appear as follows:

1. A noise is coming from the accessory drive on the front of a gasoline engine. Technician A says the serpentine belt can be removed to help determine if it is the source of the noise. Technician B says water dripped on the belt can help determine if the belt is the source of the noise. Who is correct?

    A. A only
    B. B only
    C. Both A and B
    D. Neither A nor B

Answer A is incorrect. Technician B is also correct.

Answer B is incorrect. Technician A is also correct.

Answer C is correct. Both Technicians are correct. The belt can be removed and the engine run for a short time to determine if the belt is the source of the noise. If the noise remains, the problem is not the belt or any accessories that the belt operates. Also, if the noise disappears when a few drops of water are put on the running belt, the technician knows the belt is the source of the noise.

Answer D is incorrect. Both Technicians are correct.

EXCEPT QUESTIONS

Another type of question form used on the ASE exams contains answer choices that are all correct except for one. To help easily identify this type of question, whenever they are presented in an exam, the word "EXCEPT" will always be displayed in capital letters. With this type of question, the one incorrect answer choice will actually be counted as the correct answer for that question. Be careful to read these question types slowly and thoroughly; otherwise, you may overlook what the question is actually asking and answer the question by selecting the first correct statement.

An example of this type of question would appear as follows:

1. A gasoline engine is being checked for an engine oil leak. All of the following could be used to help locate the source of the leak EXCEPT:

    A. Blacklight
    B. White powder
    C. Vacuum gauge
    D. Oil dye

Answer A is incorrect. A blacklight can be used to help locate the source of a leak.

Answer B is incorrect. White powder can be used to help locate the source of a leak.

Answer C is correct. A vacuum gauge is not used to locate engine oil leaks. A vacuum gauge is used to measure intake manifold vacuum or air inlet restriction.

Answer D is incorrect. Oil dye can be used to help locate an engine oil leak.

LEAST LIKELY QUESTIONS

For this type of question style, look for the answer choice that would be the LEAST LIKELY cause of the described situation. To help easily identify this type of question, whenever they are presented in an exam, the words "LEAST LIKELY" will always be displayed in capital letters. Read the entire question carefully before choosing your answer.

An example of this type of question would appear as follows:

1. A vehicle equipped with a gasoline engine overheats when pulling a trailer. Which of the following would be the LEAST LIKELY cause?

    A. Slipping fan clutch
    B. Seized fan clutch
    C. Restricted air conditioning condenser
    D. Restricted radiator

Answer A is incorrect. A slipping fan clutch may not fully engage and would fail to provide sufficient airflow across the radiator to keep the engine cool.

Answer B is correct. A seized fan clutch would run all the time; this may cause a low power complaint, but would not cause the engine to overheat.

Answer C is incorrect. A restricted air conditioning condenser would also restrict the airflow across the radiator. This could result in an engine overheating condition.

Answer D is incorrect. A restricted radiator could result in an overheated engine.

SUMMARY

The question styles outlined in this section are the only ones you will encounter on any ASE certification exam. ASE does not use any other types of question styles, such as fill-in-the-blank, true/false, word-matching, or essay. ASE also will not require you to draw diagrams or sketches to support any of your answer selections. If a formula or chart is required to answer a question, it will be provided for you.

Task List Overview

INTRODUCTION

This section of the book outlines the content areas or *task list* for this specific certification exam, along with a written overview of the content covered in the exam.

The task list describes the actual knowledge and skills necessary for a technician to successfully perform the work associated with each skill area. This task list is the fundamental guideline you should use to understand what areas you can to expect to be tested on, as well as how each individual area is weighted to include the approximate number of questions you can expect to be given for that area during the ASE certification exam. It is important to note that the number of exam questions for a particular area is to be used as a guideline only. ASE advises that the questions on the exam may not equal the number of specifically listed on the task list. The task lists are specifically designed to tell you what ASE expects you to know how to do and to help prepare you to be tested.

Similar to the role this task list will play in regard to the actual ASE exam, Delmar, Cengage Learning has developed six preparation exams, located in Section 5 of this book, using this task list as a guide. It is important to note that although both ASE and Delmar, Cengage Learning use the same task list as a guideline for creating these test questions, none of the test questions you will see in this book will be found in the actual, live ASE exams. This is true for any test preparatory material you use. Real exam questions are *only* visible during the actual ASE exams.

Task List at a Glance

The Gasoline Engines (T1) task list focuses on six core areas, and you can expect to be asked a total of approximately 50 questions on your certification exam, broken out as outlined:

- A. General Engine Diagnosis (14 questions)
- B. Cylinder Head and Valve Train Diagnosis and Repair (4 questions)
- C. Engine Block Diagnosis and Repair (4 questions)
- D. Lubrication and Cooling Systems Diagnosis and Repair (3 questions)
- E. Ignition System Diagnosis and Repair (6 questions)
- F. Fuel, Air Induction, and Exhaust Systems Diagnosis and Repair (6 questions)
- G. Emissions Control Systems Diagnosis and Repair (5 questions)
- H. Computerized Engine Controls Diagnosis and Repair (8 questions)

Based upon this information, the following is a general guideline demonstrating which areas will have the most focus on the actual certification exam. This data may help you prioritize your time when preparing for the exam.

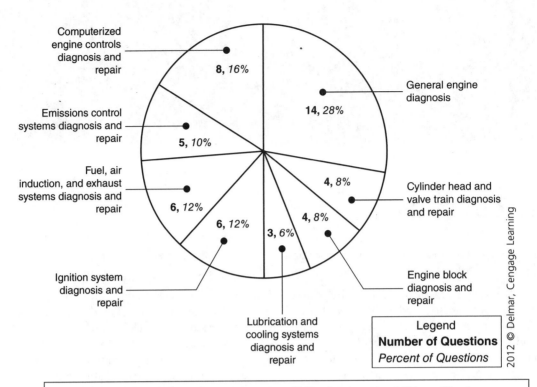

Computerized engine controls diagnosis and repair — **8**, *16%*

Emissions control systems diagnosis and repair — **5**, *10%*

Fuel, air induction, and exhaust systems diagnosis and repair — **6**, *12%*

Ignition system diagnosis and repair — **6**, *12%*

Lubrication and cooling systems diagnosis and repair — **3**, *6%*

General engine diagnosis — **14**, *28%*

Cylinder head and valve train diagnosis and repair — **4**, *8%*

Engine block diagnosis and repair — **4**, *8%*

Legend
Number of Questions
Percent of Questions

2012 © Delmar, Cengage Learning

Note: The actual number of questions you will be given on the ASE certification exam may vary slightly from the information provided in the task list, as exams may contain questions that are included for statistical research purposes only. Do not forget that your answers to these research questions will not affect your score.

A. General Diagnosis (14 questions)

1. Verify the complaint and/or road test vehicle; review driver/customer interview and past maintenance documents (if available); determine further diagnosis.

The technician must be familiar with a basic diagnostic procedure such as the following: Listen carefully to the customer's complaint, and question the customer to obtain more information regarding the problem. Identify the complaint, and road test the vehicle, if necessary. Think of possible causes of the problem. Perform diagnostic tests to locate the exact cause of the problem. Always start with the easiest, quickest test. After the repair has been made, be sure that the customer's complaint is eliminated. Road test the vehicle again if necessary.

2. Research applicable vehicle and service information, such as engine-management system operation, vehicle service history, service precautions, technical service bulletins, and service campaigns/recalls.

An important part of diagnosis is understanding the system you are working on. This applies to engine and powertrain management systems as well as other parts of the vehicle. This includes theory of operation information that is usually found in the manufacturer's vehicle service manuals or aftermarket vehicle service manuals. These manuals will also provide detailed repair information that is necessary to service vehicles, such as wiring diagrams, component replacement procedures, adjustment procedures, specifications, and test procedures.

Gather previous vehicle service information from the customer if service or repairs were performed elsewhere. In addition, access information regarding any repairs already performed by the shop, which should be kept in a customer file or on a shop management computer. Knowing what repairs have already been performed will help you eliminate duplicate services.

Further information is available in the form of manufacturers' technical service bulletins (TSBs) and recall notices. These may be in printed or electronic format, in which case the bulletins can be searched with a personal computer by vehicle as well as system or symptom. The bulletin can then be printed or viewed onscreen.

The Internet also provides many sources for gathering information. Vehicle and parts manufacturer websites are available and provide access to service information or the ability to purchase manuals and training material online.

Some hand-held equipment available today has diagnostic information embedded in the tools' software for easy technician access. This form of information access is very useful and well suited to being used with modern tools such as scan tools and labscopes or graphing multi-meters.

Lastly, vehicle repair diagnostic hotline services can be very helpful resources when you work on unfamiliar vehicles or complex problems. Various sources such as parts suppliers, tool manufacturers, and private/independent companies provide these services.

3. Inspect engine assembly for fuel, oil, coolant, and other leaks; determine needed action.

A technician must understand the basic fuel, lubricating, and cooling systems and components. The location of all possible leaks in these systems must be identified.

Fuel leaks are typically easy to find by looking for dampness and the familiar gasoline smell. Fuel lines must be replaced, not repaired.

Many times when trying to locate an engine oil leak, the engine has been leaking for an indefinite period and the engine oil has accumulated dirt. In this case, the engine must be cleaned to make a leak more visible. Another way to find an oil leak is to add a dye that reacts with an ultraviolet lamp to the engine oil.

Coolant leaks may be internal or external. Coolant leaks often come from hoses, gaskets and the water pump. A blacklight and dye are effective coolant leak detection tools.

Other leaks that the technician may find on the engine assembly include power steering fluid, automatic transmission fluid, and, possibly, brake fluid.

If a vacuum leak is sufficient enough to cause the engine to stall while idling, it should be possible to locate the leak by listening for the whistle at the source. Another method of vacuum leak detection is to spray propane around suspected areas while observing engine RPM with a tachometer. Since the propane is combustible, it will enrich the mixture, which will increase engine speed. Another popular method of finding vacuum leaks is to use a smoke generator. With the engine off, introduce smoke into the intake manifold, and then simply observe where it escapes.

4. Diagnose noises and/or vibration problems related to engine performance; determine needed action.

During engine acceleration, worn pistons and cylinders cause a rapping noise, and worn main bearings cause a thumping noise. Worn camshaft bearings usually do not cause a noise unless severely worn.

The most common noise complaint is valve train noise. The technician identifies valve tappet noise by a light, regular clicking sound that comes from the upper portion of the engine at half the engine speed. Dirty hydraulic lifters, lack of lubrication, and misadjusted valve clearance are some of the causes of valve train noise.

Ignition detonation and pre-ignition can cause noises that can be mistaken for internal engine component failures. The sound is the result of a second flame front that starts after the spark plug ignites. When the two flame fronts collide, there is a loud explosion or knock. These conditions can be caused by failure of internal engine components or by low quality fuel or a faulty ignition system. Other components that can create noise include flywheels, harmonic balancers, any belt-driven component, torque converters, and motor mounts.

5. Diagnose the cause of unusual exhaust color, odor, and sound; determine needed action.

Blue smoke is usually associated with oil. Black smoke indicates excess fuel consumption or a lack of air in the combustion chamber.

Blue-colored exhaust smoke indicates excessive oil in the combustion chamber. Blue smoke on acceleration is indicative of worn piston rings or excessive oil pressure/volume directed onto the cylinder wall area. Blue smoke at start up or deceleration usually indicates excessive wear to the valve guide/seals. Black smoke is an indication of incorrect combustion, usually a result of either restricted airflow and/or over-fueling.

White smoke indicates moisture either from condensation or coolant. White smoke during warm-up is a normal condensation of the air and is not a concern, while continuous white smoke after warm-up is usually an indication of coolant leakage.

6. Perform engine manifold vacuum or pressure tests; determine needed action.

The technician connects a vacuum gauge directly to the intake manifold to diagnose engine and related system conditions. When a vacuum gauge is connected to the intake manifold, the reading on the gauge should provide a steady reading between 17 and 22 inches of mercury (in. Hg), with the engine idling. Abnormal vacuum gauge readings indicate these problems:

- A low, steady reading indicates late ignition timing.
- If the vacuum gauge reading is steady and much lower than normal, the intake manifold has a significant leak.
- When the vacuum gauge fluctuates and reads low on a carbureted engine at idle speed, the idle mixture screws may require adjusting. On a fuel-injected engine, the injectors may require cleaning or replacing.
- Burned or leaking valves cause rapid and regular fluctuations on a vacuum gauge.

- Weak valve springs result in a vacuum gauge fluctuation.
- A leaking head gasket may cause a vacuum gauge fluctuation.
- If the valves are sticking, the vacuum gauge readings may fluctuate.
- If the vacuum gauge drops to a very low reading when you accelerate the engine and hold a steady higher RPM, the catalytic converter or other exhaust system components may be restricted.

7. Perform cylinder power balance test; determine needed action.

If the cylinder is functioning normally, a noticeable RPM decrease occurs when a cylinder misfire is created by analyzer cutout. If there is little RPM decrease when the analyzer causes a cylinder to misfire, the cylinder is not contributing to engine power. When this happens, the engine compression, ignition system, and fuel system should be checked to locate the cause of the problem. An intake manifold vacuum leak may cause a cylinder misfire with the engine idling or operating at low speed. If this problem exists, the misfire will disappear at a higher speed when the manifold vacuum decreases. When all the cylinders provide the specified RPM drop during a sequential cut-out test, the cylinders are all contributing equally to the engine power.

8. Perform cylinder compression test; determine needed action.

Disable the ignition and fuel-injection system before proceeding with a compression test. Open the throttle to wide-open throttle (WOT) position. This will allow the maximum of air into the cylinder. During the compression test, the engine is cranked through four compression strokes on each cylinder, and the compression readings are recorded. Note: Some manufacturers may recommend more or less than four strokes. In this case, the manufacturer's recommendations should be followed. Interpret lower than specified compression readings as follows:

- Low compression readings on one or more cylinders indicate worn rings or valves, a blown head gasket, or a cracked cylinder head.
- A gradual buildup on the four compression readings on each stroke indicates worn rings, whereas little buildup on the four compression strokes is usually the result of a burned exhaust valve.
- When compression pressure readings on all cylinders are equal but lower than specified, suspect worn rings and cylinders or a jumped timing chain.
- A leaking head gasket or cracked cylinder head can cause low compression on two adjacent cylinders.
- A hole in the piston or a severely burned exhaust valve can result in zero compression in a cylinder. If the zero compression reading is caused by a hole in the piston, the engine will have excessive blowby.

Higher than specified compression pressure usually indicates carbon deposits in the combustion chamber. Higher compression can cause spark knock and detonantion, which will result in internal engine damage. On the newest engines, the electronic control module (ECM) will sense the spark knock and retard the ignition timing to compensate. Retarding the ignition timing will reduce engine power.

9. Perform cylinder leakage/leak down test; determine needed action.

During a cylinder leakage test, regulated shop air is ported into the cylinder at top dead center (TDC). Move the piston of the cylinder to be tested to TDC on the compression stroke with both exhaust and intake valves closed. The gauge on the leakage tester indicates the percentage of air escaping the cylinder. A gauge reading of 0 percent indicates no cylinder leakage; if the reading is 100 percent, the cylinder is not sealing at all.

An excessive reading is one that exceeds 20 percent. Check for air escaping from the tailpipe, the positive crankcase ventilation (PCV) valve opening, or the top of the throttle body or carburetor. Air escaping from the tailpipe indicates an exhaust valve leak. Air escaping out of the PCV valve opening indicates that the piston rings are leaking. An intake valve is leaking if air is escaping from the top of the throttle body or carburetor. Remove the radiator cap and check the coolant for bubbles, which indicate a leaking head gasket or cracked cylinder head.

10. Diagnose engine mechanical, electrical, electronic, fuel, and ignition problems with an oscilloscope, engine analyzer, digital multi-meter (DMM), and/or scan tool; determine needed action.

Oscilloscopes are used to measure voltage while looking at a time scale. They are very useful when diagnosing faulty sensors or ignition systems. Oscilloscopes are often used when trying to find intermittent conditions or sensor problems that have not set diagnostic trouble codes (DTCs).

On many engine analyzers, ignition performance tests include primary circuit tests, secondary voltage/firing voltage (kV) tests, acceleration tests, scope patterns, and cylinder miss recall. Primary circuit tests include coil-input voltage, coil primary resistance, dwell, curb idle speed, and idle vacuum. The firing voltage test measures the voltage required to fire a spark plug. A high resistance problem in a spark plug or a spark plug wire causes higher firing voltage, whereas a fouled spark plug or a cylinder with low compression results in a lower firing voltage.

Some firing voltage tests include a snap voltage test in which the analyzer directs the technician to accelerate the engine suddenly. When this action is taken, the firing voltage should increase evenly on each cylinder. Some engine analyzers also display circuit gap for each cylinder. The circuit gap is the voltage required to fire all the gaps in the secondary circuit, such as the rotor gap, but the spark plug gap is excluded. Some analyzers display the burn time for each cylinder with the firing voltage test. The burn time is the length of the spark line in milliseconds (ms). The average burn time should be 1 to 1.5 ms.

DMMs are critical to working on engine electrical components. As tools for checking voltage, amperage, and resistance, they are indispensable to the technician. They are used to diagnose a circuit fault, usually after a DTC has been set for a circuit.

A scan tool is a must for diagnosing an engine concern. It is now standard practice to connect this tool to the ECM and check for DTCs prior to starting any repair. The self diagnostics available to the technician through the ECM can save valuable time when diagnosing a concern.

11. Inspect engine compartment wiring harness, connectors, seals, locks, vacuum hoses; check for proper routing and condition; determine needed repairs.

The first step in troubleshooting a malfunctioning engine is to visually inspect the engine compartment for any unsecured electrical connection or for apparent corrosion. Also check for any leaking or restricted (pinched) vacuum lines. Next, wiggle all related connections and lines while the engine is running and observe any changes in the engine's operation.

12. Observe and interpret instrument panel gauge readings.

It is advisable to compare the readings of the dash panel gauges with a known accurate shop meter to verify the accuracy of these units. Then, compare these readings to the manufacturer's specifications.

13. Read and interpret electrical schematic diagrams and symbols.

Electrical schematics, also known as *electrical diagrams* or *wiring diagrams*, are the road maps technicians use to properly diagnose and understand the vehicle's electrical circuits. By studying and completely understanding an electrical schematic, many technicians are able to identify electrical problems with no other diagnostic information. Understanding the symbols used on the schematics is a major key to fully utilizing the diagrams as a diagnostic tool. You must be able to identify where the circuit power originates, whether the device is voltage-controlled or ground-controlled, and what other devices are powered from the same source. Identifying terminal pin numbers, wire colors, which fuse(s) feed the circuit and their location, circuit identifying numbers, splices, connectors, and where a circuit is grounded are all quite necessary in nearly every type of electrical operation, and accurately identifying them is possible if you have access to the proper wiring schematic.

Most diagnostic procedures are written with the assumption that the technician performing the test is familiar with the location and types of circuits being tested. Having a specific electrical schematic of that circuit before you, and an ability to gain information from the schematic, will ensure that you can identify what is being tested with the most efficiency.

Many computer- or Internet-based information systems provide the ability to print out complete schematics or even to zoom in on a specific portion of it and print the zoomed image. The technician can then write notes on the printed copy or even highlight current flow or other pertinent items that need to be focused on.

14. Test and diagnose emissions or driveability problems caused by battery condition, connections, or excessive key-off battery drain; determine needed repairs.

Battery voltage is not only needed to start the engine but is also very important in stabilizing the voltage during engine operation. Low battery voltage or state of charge, as well as poor battery cable connections, will cause slow engine cranking and hard-start or no-start complaints. Once the engine is running, the vehicle's charging system supplies the voltage needed and restores the battery's charge. As the charge voltage is supplied to the system, it will fluctuate depending on vehicle electrical loads and the sensed need of the battery. Once the battery has the same voltage as the output of the charging system, the system voltage and charging output are leveled.

Many vehicle's systems are monitored by electronic components that send information to a control module or computer. This information is delivered as changes in voltage. Therefore, precise voltage control is important to effective engine control management. When there is a large fluctuation in system voltage or voltage spikes, the computer may reset base sensor input levels stored in its memory, and drivability problems may result.

Most computers have a few milliamperes of current draw when they are not in operation. This current draw is called parasitic load. Since many vehicles have several computers, this current draw may discharge a battery if the vehicle is not driven for several weeks. Most manufacturers allow between 50 to 100 milliamps of parasitic current draw. A value greater than 100 milliamps is considered excessive, and further diagnostics should be performed to determine what is causing the current drain.

Connecting a DMM set to measure amperage in series between the negative battery cable and the battery post will allow you to measure this load. All accessories must be off and the key removed from the ignition. Some vehicles may require a long wait period of up to one hour before all computers will power down. Take a reading and compare it to manufacturer's specifications.

15. Perform starter current draw test; determine needed action.

Starter current draw testing is performed only on batteries with an open circuit voltage reading of 12.6 volts or greater. Several different testers can be used. If an analog tester is used, always check the mechanical zero on each meter and adjust as necessary. Be sure that all electrical loads are off and the doors are closed, as additional loads will cause additional draw. The ignition is disabled and the engine is cranked while you observe the ammeter and voltmeter readings. High current draw and low cranking speed usually indicate a defective starter. High current draw may also be caused by internal engine problems. A low cranking speed and low current draw with high cranking voltage usually indicate excessive resistance in the starter circuit, such as in the cables and connections.

16. Perform starter and charging circuit voltage drop tests; determine needed action.

The resistance in an electrical wire may be checked by measuring the voltage drop across the wire with normal current flow in the wire. To measure the voltage drop across the positive battery cable, connect the positive voltmeter lead to the positive cable at the battery, and connect the negative voltmeter lead to the other end of the positive battery cable at the starter solenoid. Disable the ignition system. Crank the engine. The voltage drop indicated on the meter should not exceed 0.5 V. If the voltage reading is above this figure, the cable has excessive resistance. If the cable ends are clean and tight, replace the cable.

To measure the voltage drop across the starter solenoid, connect the positive voltmeter lead to the positive battery cable on the starter solenoid, and connect the negative voltmeter lead to the starting motor terminal on the other side of the solenoid. Leave the voltmeter on the lowest scale and crank the engine. If the voltage drop exceeds 0.3 V, the solenoid disc and terminals have excessive resistance.

To measure the voltage drop across the starter ground circuit, connect the positive voltmeter lead to the starter motor housing and the negative lead to the negative battery post and crank the engine. If the reading is greater than 0.2 V and the cable connections are clean and tight, replace the negative battery cable.

17. Inspect, test, and repair or replace components, connectors, and wires in the starter and charging control circuits.

The starter and charging system circuit diagnosis should begin with a visual inspection followed by the technician grabbing and gently moving the wiring harness and connectors while looking for insulation damage and loose connections.

When testing the starter control circuit, connect the positive voltmeter lead to the positive battery cable at the battery, and connect the negative voltmeter lead to the solenoid winding terminal on the solenoid. Leave the ignition systems disabled. If the voltage drop across the control circuit exceeds 1.5 V while the engine is cranked, individual voltage drop tests on control circuit components are necessary to locate the high resistance problem.

A similar test can be performed on the control circuit of the charging system to check for voltage drop. While performing this test on the charging system, the engine must be running and the charging system operating under load. The charging system can be put under load by applying a load to the battery.

Wiring harnesses can usually be repaired using proper soldering techniques. Battery cables are typically replaced. However, it is acceptable to replace the terminal end if necessary. If the terminal end is replaced, care should be taken to ensure that the cable is cut back far enough to remove all corrosion. Alternators and starters are typically replaced, not repaired. When mounting the new unit, ensue that the fasteners are torqued correctly.

18. Differentiate between electrical and mechanical problems that cause a slow crank, no crank, extended cranking, or a cranking noise condition.

If the starter fails to crank the engine, the problem may range from a faulty starter motor to broken components inside the engine. If no sounds come from the starter motor when it is activated, first disable the ignition system. Then, attempt to rotate the crankshaft pulley by hand in the normal direction of rotation. If the crankshaft can be rotated freely through two complete revolutions, then diagnosis of the vehicle's starting system is the next step.

If you are unable to rotate the crankshaft by hand, the engine may be hydrostatically locked or have broken internal components. To check for hydrostatic lock, remove all the spark plugs and attempt to rotate the crankshaft again. If oil or coolant squirts from the spark plug holes, this indicates a bad head gasket, a warped cylinder head or block, a cracked cylinder head or block, or that the customer drove through deep water. Water squirting from the spark plug holes could also be an indication that the customer drove through deep water, drawing water in through the air intake.

If the crankshaft cannot be rotated at all with the spark plugs removed, or it cannot be rotated through at least one complete revolution, the engine may be seized or have broken internal parts. Pull the dipstick and check crankcase oil level. If oil does not register on the dipstick, it is possible that the pistons are seized in their bores or that the connecting rods are seized to the crankshaft. If the oil level is sufficient, a broken component may have lodged between moving parts inside the cylinder block, preventing the parts from rotating.

Many engines are non-freewheeling or "interference" engines. On these engines, a no-crank condition may be caused by piston-to-valve contact. This is a common occurrence when a timing belt slips or breaks, but it may also occur on engines fitted with a timing chain and sprockets. On many OHC engines, it is possible to easily loosen or remove part of the timing belt cover. Do this, if possible, and check for obvious signs of cam drive failure.

If the customer states that the starter cranks the engine but it will not start (or takes a long time to start), confirm that the valve train is operating properly before attempting to crank the engine your self. If the timing belt or chain is broken or jumped, additional cranking may cause severe engine damage.

A no-start or hard-starting complaint can be caused by a faulty ignition, fuel emission control systems, or severely worn internal engine components. These complaints can also be caused by broken or slipped valve train timing components, especially on free-wheeling engines. A broken timing device may cause some cylinders to have good compression while others have none. A slipped timing device may result in all cylinders having low compression. To determine if the belt or chain is functioning properly, rotate the crankshaft by hand while observing the distributor rotor or camshaft. If these components fail to rotate with the crankshaft, the timing belt or chain is broken. It is common for timing belts that have not been replaced at OEM intervals to have their teeth torn off by the crank sprocket. The timing belt will appear normal during the visual inspection but fail to rotate as the engine is turned over. If the belt does rotate with the crankshaft, confirm proper indexing of the rotor or camshaft to determine if the belt or chain has slipped. Rotate the crankshaft until the piston in cylinder #1 is at TDC on the compression stroke. Then, check the distributor rotor or camshaft position to make sure that it is correct.

During a no-start diagnosis, most technicians will check for the presence of spark before checking for fuel or mechanical issues. The process for checking spark requires using a tool called a *spark tester* in place of one of the spark plugs and cranking the engine to visually check for spark. Some distributorless ignition systems (DIS) are not compatible with this type of test and may be tested using appropriate secondary ignition lab scope procedures.

Fuel-injected engines usually receive high-pressure fuel from an electric pump. To verify that the fuel pump is operating and fuel is reaching the engine, locate the fuel line that supplies fuel to the throttle body or fuel injector rail. Turn the key on/engine off after installing a fuel pressure gauge. Verify that the gauge registers adequate pressure. If the gauge does not register any pressure or registers very low pressure, proceed with fuel system diagnosis. If fuel pressure is adequate, begin diagnosis of the fuel injection control system.

19. Test and diagnose engine performance problems resulting from an undercharge, over-charge, or a no-charge condition; determine needed action.

The charging system is responsible for maintaining stable electrical system voltage. Undercharging as well as over-charging may cause engine performance problems, including hard-starting or battery gassing and corrosion concerns. A defective diode in the alternator may allow enough AC voltage leakage into the electrical system to disrupt normal computer operation. Test the alternator with a DMM or labscope for excessive AC voltage output and replace it if found out of specifications.

If alternator output is zero, the alternator field circuit may be open. The most likely place for an open circuit in the alternator is at the slip rings and brushes. If alternator output is normal but no charging current is measured at the battery, the fuse link between the alternator output terminal and positive battery cable is probably open. If the alternator output is less than specified, always be sure that the belt and belt tensions are satisfactory. If belt condition and tension are satisfactory and the alternator output is less than specified, the alternator is defective. On some alternators, there is a method of "full fielding" the unit. This technique will by-pass the voltage regulator circuit, and full alternator output will be obtained. In this case, if the alternator has full output, then the regulator or its circuit has failed.

20. Inspect, adjust, and replace alternator (generator) drive belts, pulleys, tensioners, and fans.

A loose belt causes low alternator output and a discharged battery. A loose, dry, or worn belt may cause squealing and chirping noises during acceleration and cornering. Belt tension may be checked by measuring the belt deflection. Press on the belt with the engine stopped to measure the belt deflection. One half inch (12.7 mm) per foot (30.5 cm) of free span is usually acceptable.

Serpentine drive belt systems will most often use automatic belt tensioners. Automatic tensioners are usually spring loaded, but they should be checked to be sure they are applying adequate pressure to the belt to prevent slipping or squealing from the belt. Pulley alignment is critical on serpentine belt drives. Any misalignment will cause noise from the belt and may allow the belt to slip off the pulleys.

Whenever a customer with a serpentine drive belt engine consistently complains of belt noise, you should check to see if any previous service has been performed on the engine that may have required accessories to be removed. If a pulley is reinstalled backward or any spacers are left off accessory brackets, pulley misalignment can occur and cause constant belt noise. Proper routing of the serpentine belt must be observed when replacing a drive belt. Routing the belt incorrectly can cause some accessories to spin backward or not allow proper tension to be applied to the belt by the belt tensioner.

B. Cylinder Head and Valve Train Diagnosis and Repair (4 questions)

1. Remove, inspect, disassemble, and clean cylinder head assembly(s).

Remove a cylinder head only when the engine is cold. Removing a warm cylinder head may cause the head to warp, especially if it is made of aluminum.

Remove the cylinder head bolts, loosening the bolts in a sequence opposite that of the tightening sequence. Note and record the positions of special bolts. Remove the cylinder head from the engine. Cylinder heads can be quite heavy, so ask an assistant to help you, especially if the engine is still mounted in the vehicle.

Use a spring compressor to compress the valve springs and then remove the valve locks, or *keepers*. Release the compressor and remove the retainer, rotator, spring, and spring seats from the head. Keep all parts in an organizer so they can be returned to their original cylinder. Check the valve stem tips for mushrooming. If it is present, the tip must be dressed with a file before the valves are removed from the head. Remove the valves from the cylinder head and place them in an organizer.

When you remove the cylinder head from an overhead camshaft (OHC) design engine, the timing belt or chain must first be disconnected from the camshaft. The procedure for doing this varies from one manufacturer to another. On some engines with a chain-driven camshaft, the camshaft sprocket is unbolted from the cam. The cylinder head assembly is then removed, leaving the chain and sprockets in position on the engine. On some engines with a belt-driven camshaft, the belt tensioner is loosened and the belt is slipped off the camshaft sprocket. On other engines, the timing cover and belt must be completely removed from the engine. If the timing belt is being removed from the engine and will be reused, mark the direction of rotation on the belt. Reinstall the belt so it rotates in the same direction. Never crank the engine after a timing device has been loosened or removed (until the cylinder head has been removed). Cranking an engine while the timing belt is loose or disconnected can cause immediate and serious engine damage.

The basic cylinder head removal procedure varies from one manufacturer to another. On some OHC engines, remove the cylinder head, camshaft(s), and rocker arms (if used) as an assembly after loosening and removing the cylinder head bolts. On some engines, the rocker shaft and arm assembly (including the upper half of the cam bearings) must first be removed to access cylinder head mounting bolts. Refer to the appropriate service manual for information.

When inspecting the cylinder head, it is normal if carbon buildup is limited to a light, even layer across the entire combustion chamber. If the carbon is excessive, the cylinder head should be thoroughly cleaned. Excessive carbon buildup could be caused by worn valve guides, valve seals, or rings.

Cleaning the components for reassembly changed dramatically with the advent of aluminum heads and blocks. Many of the surfaces of these components are polished to near mirror-like qualities. After you clean these components with the appropriate degreasing solution, carbon buildup may be removed by hand with a soft wire wheel or in an abrasive cabinet like a glass blaster. In either case, care must be taken to ensure that gasket surfaces remain in their original condition. Use of refinishing wheels on polished surfaces can cause new gaskets to fail prematurely. The head must be carefully washed after removing deposits and old gasket material to ensure that none of the debris restricts passages. Many manufacturers recommend plastic scrapers and gasket-softening chemicals to remove old gaskets. Be sure to find out what the manufacturer recommends to guarantee a quality repair.

2. Inspect threaded holes, studs, and bolts for serviceability; service/replace as needed.

When removing or installing a component, carefully inspect all bolts, studs, holes, and nuts for damage to threads. Stripped, cross-threaded, nicked, rolled, and rusty threads can produce errors in the torque values. Use compressed air to remove foreign matter from holes. Obstructions lodged in holes, such as metal and liquids, can cause a bolt to bottom out and reach torque value without obtaining the required clamping force. Examine the shank of bolts and studs for signs of twisting or over-torque fracturing. Replace defective bolts and studs. Ensure that nuts are not distorted or cracked, and replace self-locking and defective nuts. Clean and dress threads in holes. If threads are damaged beyond use, drill and tap the hole and install an approved HeliCoil. Use an approved stud extractor to remove broken studs and bolts.

3. Measure cylinder head thickness; check mating surfaces for warpage and surface finish; inspect for cracks/damage; check condition of passages; inspect core and gallery plugs; determine serviceability and needed repairs.

Cylinder head thickness is measured from the head gasket surface to the valve cover gasket surface. If the head thickness is less than specification, it is an indication that the head was previously milled. A head that is less than specification must be replaced.

Inspect the cylinder head for cracks and other obvious damage. Remember, not all cracks are visible to the eye. Therefore, it may be necessary to perform additional tests if a crack is suspected. Common locations for cracks are between the valve seats or around the spark plug hole.

Look for cracks between adjacent valves. On engines with two exhaust valves per cylinder, check carefully for cracks between adjacent exhaust valves. Cracks may be found using magnetic particle inspection, a dye penetrant, or by pressure testing in a tank.

If the cylinder head has excessive damage, it may be more cost efficient to replace it instead of repairing it.

Check the cylinder head mating surface for finish and warpage. Surface finish can be checked using a surface comparator gauge. Use a straightedge with a feeler gauge to measure warpage, which can occur in any direction on the head surface. Measure for warpage in three areas along the edges and three areas across the center. Compare the measurements with the manufacturer's specifications. If there are no specifications available, a rule of thumb is 0.003 in. (0.08 mm) for any six-inch length. Use the feeler gauge to determine whether the cylinder head needs to be resurfaced.

Use the straightedge in the same manner to check for warpage of the intake and exhaust manifold mating surfaces. The general rule for maximum warpage limit for the manifold mating surfaces is 0.004 in. (0.1 mm).

4. Inspect valves, guides, seats, springs, retainers, locks, and seals; determine serviceability and needed repairs.

The valves should be carefully examined not only to determine their suitability for continued use, but also to determine the cause of the failure so that a repeat failure can be avoided. Check valves for the following conditions:

Valve stem wear: Measure the valve stem at the area of greatest wear and compare it to specifications. Examine the surface of the valve stem where it is worn for galling or scoring caused by worn valve guides or carbon deposits. Check the tip of the stem for wear or flattening, which may be the result of improper rocker arm alignment or excessive valve lash. Check for a condition known as *necking*, a narrowing of the stem just above the fillet. Valves having stems that are worn, scored, galled, necked, or have stress cracks should be replaced. The condition causing the failure should be corrected.

Valve head: Check the head of the valve for the presence of cupping, indicating that it has been subjected to excessive heating. Valves that are cupped should be replaced. Check the valve face for pitting, channeling, cracking, or burning. Small amounts of pitting and wear can be corrected by resurfacing. Some manufacturers specify an interference angle of one degree between the valve and seat to break up and reduce carbon deposits. After the valve is resurfaced, check the margin. If the valve cannot be resurfaced without reducing the margin below specifications, it should be replaced.

It is very important that the valve make good contact with the valve seat, because heat that the valve is subjected to must be transferred to the cooling system. Improper seating due to worn valve guides, weak springs, or carbon buildup will shorten the life of the valves.

Before attempting to measure valve guide wear and to measure valve seat runout, it is important to use a valve guide cleaner or a bore brush. After preparing the guide, use a small bore gauge and measure the valve guide at three different locations from top to bottom. Measure the fingers at each different location of the bore gauge with an outside micrometer. Measure the diameter of the valve stem, and subtract the difference to find the clearance. Another technique is the *valve rock* method. Insert the correct valve into the guide and install the special tool that maintains the height of the valve during the inspection. Place a dial indicator on the cylinder head with the tip of the dial indicator at a right angle to the valve head. Zero the dial indicator. Rock the valve and observe the clearance indicated on the dial.

The maximum wear limit that most manufacturers recommend is 0.005 in. (0.12 mm) or less. Desirable clearances are typically 0.001 to 0.003 in. (0.025 to 0.080 mm) for intake and 0.0015 to 0.0035 in. (0.04 to 0.09 mm) for exhaust valves. If any of the valve guides are out of specification, replace the guide.

The valve seat will first have to be checked for wear and cracks. After performing a visual inspection, a thorough examination for cracks using the appropriate equipment is recommended, as minute cracks will be undetectable to the human eye. Valve seat angle must then be confirmed by resurfacing with either a tool bit or grinding stone method. To check for concentricity, a new or resurfaced valve can be installed into the guide and seated with a light tapping motion, removed, and then inspected. A light circular mark on the valve seat of the valve indicates that the interference angle between the valve and cylinder head is correct and that the valve is centering properly.

Valve spring retainers and locks must be checked for wear and scoring. Check for signs of cracks and areas of discoloration. When any of these conditions are present, replace the components. The valve lock grooves on the valve stems must be inspected for wear, particularly round shoulders. If these shoulders are uneven or rounded, replace the valve. It is a good practice to replace the valve stem seals any time the cylinder head is disassembled.

Check the valve springs for squareness using a square. Valve springs that are out of square will result in side loading of the valve stem and will lead to rapid wear. Measure the valve spring length and then test for correct spring pressure at the specified height and compare to specifications. Since the valve springs are crucial to closing the valve rapidly and maintaining pressure while seated, you should replace any weak or defective springs.

When resurfacing a valve seat, remove only enough material to restore the valve seat to recommendations. Make sure the cutting tool is of the correct angle and is in good condition. Resurface stones before beginning valve seat resurfacing to ensure good results, and only apply as much force as necessary against the stone. A valve seat that is too wide after resurfacing will need to be narrowed using grinding stones with different angles. If it cannot be restored in this manner, the seat may need replacement.

Valve seals are replaced when a cylinder head is reassembled. However, inspecting the old seals will allow the technician to determine if the oil-consumption concern was possibly caused by worn valve guide seals. Valve guide seals that are hard and brittle can be an indication of inferior quality seals or engine overheating.

▪ 5. Reassemble, check, and install cylinder head assembly(s) and gasket(s) as specified by the manufacturer.

Cleanliness is most important when reassembling and installing the cylinder. All parts should be clean and lubricated. Before installing the cylinder, make a final check that the block deck surface is prepped and then lay the new head gasket down, making sure to orient it correctly. The head should be set down on the block using alignment dowels, and all the bolts should be lubricated and installed in their correct holes. Care should be taken to ensure head bolts that were previously installed outside the valve cover are not installed under the valve cover. These bolts will have paint on the head, and if they are installed under the valve cover, the paint can be loosened by the hot engine oil and act as an abrasive on engine components. Cylinder head bolts must be torqued in the proper sequence. Some head bolts will be torqued in a three-step process; this is done by taking the final torque value, dividing it into thirds, and tightening the head bolts in sequence three times, each time going to a higher torque value. Other styles of head bolts are torque-plus-angle. These head bolts are torqued to specification in sequence and then turned an additional number of degrees. Some OEMs specify that these head bolts must always be replaced.

6. Inspect pushrods, rocker arms, rocker arm shafts, electronic wiring harness, and brackets; repair/replace as needed.

Before assembling the rocker arm and the related components, some inspection of the components should be performed. The rocker arms must be inspected for wear and damage. Rocker arms are constructed of cast iron, stamped steel, or aluminum. Three areas on the rocker arm receive high stress: the pushrod contact, the pivot, and the valve stem contact surface. The pushrod and the valve stem contact areas should be round and show signs of even wear. Oil passages in the rocker arms should be clean and free from debris.

Inspect rocker arm shafts, if applicable, for straightness and for excessive wear at the points where the rocker arm rides. The shafts should be free of scoring and galling. Inspect the shaft for wear at the places where the rocker arm shaft pivots. A slight polishing with no ridges indicates normal wear. Wear in this location usually indicates a lack of lubrication, which in turn generates excessive heat. The shaft itself can be checked for straightness by rolling it on a smooth surface.

Pushrods should be inspected for signs of wear and bending. Inspect the tips for normal wear patterns. Check the pushrod runout by rolling it on a known true surface; a piece of glass works well. Runout should not exceed 0.003 in. (0.08 mm).

Many engines now have a wiring harness under the valve cover. This harness should be visually inspected, and cuts, abrasive wear marks or lumps in the insulation will require harness replacement. All harness mounting brackets must be in place and secure. A cracked or distorted bracket must be replaced to protect against vibration damage and ensure the harness stays in the correct location during engine operation.

7. Inspect, install, and adjust valve lifters and retainers; adjust valve clearance.

When inspecting valve lifters, the surface face of the lifter must be smooth with a centered circular wear pattern. The surface should be a convex counter-machined face. If wear extends to the edge of the lifter, the convex shape is worn away, and the lifter must be replaced. The lifter body should be polished and smooth, free from any ridges, scoring, and signs of scuffing.

Hydraulic lifters that pass the visual inspection should be tested for leak down. This is done using a special tool that applies weight to a primed lifter submerged in engine oil. While the lifter is bleeding down, observe the scale and compare the rate of bleed down over a certain length of time. Leak down range is between 20 and 90 seconds.

Sometimes it might be necessary to disassemble a lifter. Only disassemble one lifter at a time. Keep track of the order in which it comes apart.

Flat tappet lifters require a break-in period. A normal break-in involves applying break-in lube on the camshaft and the lifter face. Immediately following start up, the engine must be run at varying speeds between 1,500 to 2,500 rpm over a 20-minute period. The engine is run at off idle speeds to provide adequate oil to be splashed onto the machined surfaces that are in contact with the camshaft.

The following are typical methods of adjusting valve lash:

- Adjustable nut attaching the rocker arm to the stud
- Adjustable screw located in the actuation end of the rocker
- Shims positioned between the camshaft lobes and the followers
- Pushrod length adjustment
- Measuring valve stem height and removing material from the valve tip to correct

A typical adjustment procedure requires that the cylinder to be adjusted be positioned at TDC on the compression stroke. Adjust the hydraulic lifters to the specified preload while setting the specified clearance for solid lifter engines. Usually, after the #1 cylinder is adjusted, the crankshaft must be rotated to the next designated location and the specified valve clearance adjusted. Depending on the manufacturer and number of engine cylinders, all the valves can be adjusted at two or three different crankshaft locations, while others require locating each piston to TDC.

8. Inspect, measure, and replace/reinstall overhead camshaft and bearings; measure and adjust end-play.

A careful inspection of the camshaft will ensure that proper valve and injector timing can be obtained. Because the camshaft is constantly supplied with oil, cleaning should be as simple as rinsing with dip tank solution and blowing out with compressed air. Oil passages may require cleaning with a wire brush. Inspect the cam lobes for signs of pitting, scoring, or flat spots. Discard any camshaft that does not pass a visual inspection. If the camshaft passes the visual part of the inspection, take measurements and compare with the manufacturer's specifications. Use the manufacturer's procedures for taking lobe measurements. The heel-to-toe measurement is equal to the diameter of the circular portion of the lobe and the maximum amount of lift created by the rise at the top of the lobe.

Visually inspect bearing journals for bluing, scoring, or wear. Measure the diameter of journals with an outside micrometer, and inspect for out of round condition by checking around the journal in several places. Inspect the shaft and gear keyway for cracks or distortion. Ensure the keyway in the gear and shaft matches the Woodruff key.

C. Engine Block Diagnosis and Repair (4 questions)

1. Remove, inspect, service, and install pans, covers, ventilation systems, gaskets, seals, and wear sleeves.

Inspect the oil pan and covers for cracks, dents, and areas of rust. If they are rusted, it is advisable to replace them. Check the gasket-mating surfaces for straightness using a straightedge and a feeler gauge. If the gasket surface is warped, and the pan or cover is stamped steel, it can be straightened by striking it with a ball peen hammer on a flat, true surface. Stamped-steel valve cover gasket mating surfaces can be straightened the same way that oil pan surfaces are aligned.

When installing new gaskets, the gasket and the surface that it is going to seal should be free from debris, oil, and grease. It is imperative that all of the old gasket material is removed for proper sealing.

Wear sleeves are typically replaced with the seals. Inspection of the old wear sleeves may indicate a component that has been incorrectly installed or run out of alignment.

2. Disassemble, clean and inspect engine block for cracks; check mating surfaces and related components for damage or warpage and surface finish; check deck height; check condition of passages, core, and gallery plugs; inspect threaded holes, studs, dowel pins and bolts for serviceability; service/replace as needed.

Cracks in the cylinder block are usually identified during a visual inspection. When cracks are found, attempt to determine the cause. The following usually cause cracks:

- Fatigue
- Excessive flexing
- Impact damage
- Extreme temperature changes in a short period of time
- Detonation

Inspect all the oil passages with a shop light along with an oil gallery brush and compressed air. Oil passages should be free from any gasket material, metal shavings, and other foreign objects.

Visually inspect the cylinder block deck for scoring, corrosion, cracks, and nicks. If a scratch in the deck is deep enough to catch on a fingernail, the deck needs to be resurfaced. Measure deck warpage with a precision straightedge and feeler gauge. To obtain proper results, the deck must be perfectly clean. Check for warpage across the four edges of the deck. The thickest feeler gauge that will fit between the straightedge and the deck determines the amount of warpage. If a deck is warped, you must then determine the amount of material to be removed. On V-type engine blocks, when one deck is warped and needs to be machined, both sides should be machined. Surface finish is checked using a surface comparator. Deck height is the distance from the crankshaft center line to the top of the deck surface. This height is changed when the block deck is milled. If the deck height is lower than specification, then compression will be higher than normal. Blocks with low deck height can be built up; however, block replacement is usually more cost effective.

All engine block passages should be checked for contamination. Restricted passages can prevent oil or coolant flow and result in failed components. Core and gallery plugs are replaced. Threaded holes, studs and bolts need to be checked for pitting or erosion. If there is evidence of damage, these must be replaced. Dowel pins should fit tightly in the block.

3. Inspect and measure cylinder walls for wear and damage; determine serviceability and needed repairs.

After visually inspecting the cylinder bores, use a dial bore gauge, an inside micrometer, or a telescoping gauge to measure the bore diameter. Piston movement in the cylinder bores produces uneven wear throughout the cylinder. The cylinder wears most at 90 degrees to the piston pin and in the area of the upper ring contact at TDC. This is because the cylinder receives less lubrication while being subjected to the greatest pressure and heat at the top of the ring belt.

Taper in the cylinder bore causes piston ring gap to change as the piston travels in the bore. To measure taper and out of round using a dial bore gauge, simply rotate and move the gauge up and down in the bore. If you use a telescoping gauge or inside micrometer, measure the top of the bore just below the deck and at the bottom of the ring travel.

Cylinder taper is the difference between the cylinder diameter at the top of the ring travel and cylinder diameter at the bottom of ring travel. As a rule of thumb, cylinder taper should not exceed 0.005 in., 0.127mm. (this specification varies widely, occasionally being as low as 0.001 in., 0.0254 mm) before it is necessary to have the block bored. Cylinder out of round is the difference between two cylinder diameter measurements taken 90 degrees apart. Take the first one across the thrust bore diameter and the other one along the axial cylinder bore diameter at the same height in the cylinder. Take out of round measurements at the top, middle, and bottom of ring travel areas, because cylinders wear unevenly. Maximum out of round should not exceed 0.001 in., 0.0254mm. If any of the cylinders are out of specifications, the block should be reconditioned by having all cylinders bored to a standard oversize diameter.

If the cylinder needs to be bored, determine the oversize diameter. Use the cylinder that has the most wear as a reference for overbore size. It is a good practice to match the pistons to the bore. Pistons will have some differences in size, due to manufacturing tolerances. Measure the exact size of the replacement pistons. Then, determine the desired finished size of the cylinder and how much will be bored, leaving a small amount for finishing by honing. After the cylinders have been finish honed, they should be carefully cleaned using hot soapy water and then oiled to prevent flash rust.

4. Inspect in-block camshaft bearings for wear and damage; replace as needed.

Inspect the camshaft bearings for scoring, roughness, and wear. Camshaft bearings or bearing bores should be measured at a minimum of two different locations with a telescoping gauge or dial bore gauge. Measure the camshaft journals with a micrometer, and subtract the journal diameter from the bearing diameter to obtain the clearance. If the wear exceeds specifications, replace the bearings.

The type of tool needed to remove and install the camshaft bearings depends upon the engine design. Most overhead valve (OHV), camshaft-in-block engines will use a camshaft bushing driver and hammer. The right size mandrel is selected to fit the bearing. Turning the handle tightens the mandrel against the camshaft bearing. Then, the bearing is driven out by hammer blows. The same tool is used to replace the bearings.

When installing cam bearings, it is very important that the bearing insert be properly positioned in the bearing bore. Be absolutely sure that any oil hole(s) in the bearing insert align with oil supply passages in the bearing bore. This may mean that the insert is positioned toward the front of the bore, the back of the bore, or even the center of the bore. The position is not important as long as the oil holes line up.

5. Inspect, measure, and replace/reinstall in-block camshaft; measure/and correct end-play; inspect, replace/reinstall, and adjust valve lifters.

Begin the camshaft inspection with a visual inspection of the lobes and journals. Both must be free of scoring and galling. Normal lobe wear pattern is slightly off center with a wider wear pattern at the nose than at the heel. Off-center wear is a result of the slight taper of the lobe used in conjunction with the convex shape of the bottom of the lifter; the relationship between these two surfaces causes the lifter to rotate. If the wear pattern extends to the edges of the lobe, the lifter will not rotate. If this condition exists, the camshaft must be replaced.

To measure camshaft bearing oil clearance, use an expandable bore gauge and a micrometer to measure the inner diameter of the bearing. Then, use a micrometer to measure the expandable bore gauge. Subtract the two measurements to obtain the bearing oil clearance. Refer to the manufacturer's specifications for the proper oil clearance.

Camshaft end-play is checked using a dial indicator. Specifications will often be in the 0.002–0.008 range. Usually, the larger the cam the larger the clearance specification. End-play which is greater than specification can be caused by a worn camshaft thrust washer or worn thrust surfaces on the camshaft. If either the thrust washer or the cam is worn, it is replaced. Camshaft end-play that is less than specification is generally caused by the technician's failure to correctly reassemble the components. Excessive camshaft end-play can result in camshaft position sensor damage.

Valve lifters should be primed prior to installation per the manufacturer's recommendations. The area of the lifter where it contacts the camshaft should be coated liberally with prelube. If the valve train is adjustable, the normal procedure has the technician adjust the valves while the piston is at TDC of the compression stroke.

6. Clean and inspect crankshaft and journals for surface cracks and damage; check condition of oil passages; check passage plugs; measure journal diameters; check mounting surfaces; determine needed repairs.

The crankshaft transmits torque from the connecting rods to the drive train. These pressures and rotational forces eventually result in crankshaft wear and stress. Before reusing a crankshaft, a thorough examination is needed. Begin with a good visual check for obvious damage. Check the threads at the front and rear as well as the keyways and the pilot bearing bore. Damage found to these areas will require repair or replacement of the crankshaft. Run your fingernail across the journal surfaces to feel for nicks, scratches, and scoring. If the journal is scored, it must be polished before an accurate measurement can be made. Check the crankshaft journals for both out of roundness and taper from side to side. To check for an out of round journal, compare measurements taken 90 degrees apart around the journal. Compare measurements taken on one side of a journal to those on the other side to determine if the journal is tapered, and compare your findings with specifications. Perform a magnetic particle inspection (MPI) for stress cracks, paying particular attention to the fillet areas and the lubrication passages. Clean these passages using a small bore brush followed by a spray cleaner.

7. Diagnose piston, connecting-rod bearing, and main bearing wear patterns that indicate connecting rod and crankshaft alignment or bearing bore problems; check bearing bore and bushing condition; determine needed repairs.

A piston with a normal wear pattern will have only slight polishing of the piston skirt in a straight up-and-down pattern. A wear pattern that angles off to one side indicates a connecting rod alignment problem. Inspect the rod bearings on the suspected rod for an uneven wear pattern from side to side. When one side has more wear, it is another indication of rod misalignment. You can verify connecting rod alignment using an alignment fixture. Rods should be checked for straightness and twist; check the length and the inner diameter at both ends. Compare your measurements to specifications to determine serviceability.

The bearing bore and bushing should both be round with no more than 0.001 in. taper or out of round permitted. Bushings can be replaced. Bearing bores can be reconditioned.

8. Determine the proper select-fit components such as pistons, connecting rods, and main bearings.

Some gasoline engines will have select fit components that must be matched to the specific engine. An example is pistons. Some manufacturers will assign grades of A, B, C, and D to the engine's pistons. The difference in the diameter of these graded pistons may be less than 0.0005 in. (0.0127 mm). The technician selects the correct piston to achieve the desired piston-to-cylinder wall clearance.

9. Inspect and replace main bearings; check cap fit and bearing clearances; check and correct crankshaft end-play.

Begin with a visual inspection of the bearings. Bearings that are obviously scored or worn through multiple layers of bearing material should be replaced. Look for uneven wear patterns, indicating misalignment of the rotating members. A normal wear pattern would be a uniform dull gray surface with only minimal embedding of particles in the surface. Compare your findings with the manufacturer's recommendations. Some manufacturers publish reusability guides, which are very helpful.

There are two methods of measuring assembled clearances. The first is by comparing measurements of the journal diameter with the assembled bearing bore to determine oil clearance. The second involves assembling the rods and bearings with the inclusion of a strip of soft plastic material referred to as Plastigauge®, and then disassembling and measuring the width of the crushed plastic strip. The smaller the bearing clearance, the wider the plastic strip will be when flattened out. When using this method to measure main bearing clearances with the engine still in the vehicle, be aware that the weight of the crankshaft is resting on the lower bearing half and will affect the results. To get accurate results, the weight of the crankshaft should be supported when making this measurement in the frame.

Crankshaft end-play is measured using a dial indicator. The crankshaft is shifted all the way in one direction, the dial indicator zeroed, and the crankshaft shifted all the way in the other direction. Crankshaft end-play specifications vary with the size of the crankshaft but will generally fall between 0.002 in. and 0.0008 in. Excess crankshaft end-play can be caused by worn crankshaft thrust surfaces and worn thrust bearings. If the thrust surface is worn but smooth, an oversize thrust bearing can often be purchased to take up the clearance. An alternate method to measure end-play is to use a feeler gauge. When installing the main bearings, the back side of the bearing and the surface on which it sits should be dry and clean. The surface of the bearing that is against the crankshaft should be lubricated with clean engine oil or engine assembly lubricant as specified by the manufacturer.

10. Remove and/or replace the timing chain and gears; ensure correct timing.

Inspect the timing gear train, which includes all gears and chains that drive the camshaft. The teeth on the gears must be in good shape to maintain the timing of the intake and exhaust valves. Excess tooth wear can affect engine performance and long-term operation, as well as causing damage to the pistons, cylinders, and crank. Carefully inspect for a slight roll or lip on each gear tooth in addition to looking for chipping, pitting, and burring. If a defect is present on any gear, replace it. Some manufacturers require that the gears be changed out as a unit. Manufacturer requirements should be followed. Measure the backlash using a dial indicator and compare to specifications. Try to determine the cause of damage. If removal is necessary, generally a press or puller will be required for some of the gears. Ensure the gear is properly supported on the center of the hub next to the shaft to prevent cracking or breaking and to protect the end of the shaft from which the gear is being removed. Refer to the manufacturer's service manuals for removal of intermediate and crankshaft gears.

Before installing gears on the engine block, it is important to protect the seal area of the crankshaft in case a gear comes in contact with it. A rubber hose fitted over the crankshaft will protect it. Installation of these gears is very critical, as it will affect the engine's timing. The marks must be aligned as required.

All bushings and bearings need to be inspected for wear and binding and to ensure that they have not been overheated due to lack of lubrication. Some of the bushings are not field serviceable and therefore have to be replaced as a unit.

Timing wheels or rings need to be inspected to ensure the teeth have not been damaged from handling and the clearances for the sensors are correct.

The chain should be inspected for wear. As the chain wears, it will get longer, and this will cause incorrect valve timing and a noisy engine. Timing chain guides should be inspected and replaced if they show signs of wear.

11. Inspect, measure, or replace pistons, pins, and retainers.

Begin inspecting pistons by looking for obvious damage from preignition or detonation. These will be indicated by erosion or a hole in the piston crown area or by erosion and breakage of the ring lands. Any damage found here requires replacement of the piston and correcting the cause of the failure. Check for wear in the piston ring grooves by sliding a new ring into the groove and then inserting feeler gauges into the groove to measure clearances. A worn piston ring groove will lead to early failure and ring breakage, so the piston should be replaced. Measure the piston diameter to determine serviceability. Most manufacturers recommend that the piston skirt be measured just below the center line of the wrist pin bore. Measure the skirt diameter at 90 degrees to the center line of the wrist pin. Compare your measurements to the cylinder bore diameter to determine clearances. Measure the wrist pin bushings used with full-floating pistons and also check the condition of the retaining ring grooves for signs of wear or erosion. Replace any pistons that are no longer serviceable. Press-fit piston pins are retained in the connecting rod by interference fit. To remove or replace these pins during an engine overhaul, they may be pressed out using special fixtures, or the rod may be heated. A rod oven is used to heat the small end of the rod sufficiently enough for the piston and rod to be assembled. Rods with floating piston pins have a replaceable bushing that is pressed into place and then honed to fit the piston pin.

12. Measure piston-to-cylinder wall clearance.

Piston clearance is determined by measuring the size of the piston skirt at the manufacturer's sizing point. This measurement is subtracted from the size of the cylinder bore. If the piston clearance is not within specifications, it may be necessary to bore the cylinder to accept an oversized piston.

Since most pistons are cam ground, it is important to measure the piston diameter at the specified location. Some manufacturers require measurements across the thrust surface of the skirt center line of the piston pin boss. Others require measuring a specified distance from the bottom of the oil ring groove. Always refer to the appropriate service manual for the engine that you are servicing. An alternative method of checking piston-to-cylinder wall clearance involves installing the piston upside down into the cylinder with an appropriate sized feeler gauge sliding in beside the skirt.

13. Check ring-to-groove fit and end gaps; install rings on pistons. Assemble pistons and connecting rods and install in block; install rod bearings and check clearances.

Measure the ring groove for wear by installing a new ring backward in the groove and using a feeler gauge to measure the clearance. Check the groove at several locations around the piston. If the side clearance is excessive, replace the piston. Excessive side clearance can result in ring breakage.

The topmost ring groove wears the most. Normal ring-to-groove side clearance is between 0.002 in. and 0.004 in. (0.05 mm. to 0.10 mm). With a new ring located in the groove, slide in a feeler gauge the size of the maximum clearance specification. If the feeler gauge slides, the clearance is excessive. Locating the new ring into the ring groove in the same manner as checking clearance can also check the depth of the groove. Roll the ring around the entire groove while observing for binding. The ring depth should remain consistent. To check ring gap, place the ring into the appropriate cylinder bore. Use an inverted piston to slide the ring to the specified depth, usually the bottom of ring travel in the stroke. The inverted piston crown will keep the ring square in the bore. Measure the gap at the ring ends. The value will be in the range of 0.004 in. per inch of diameter.

When installing rings on the pistons, compression rings should be installed using a ring expander, and oil control rings can be spiraled on if they are of the three-piece design. Ring end gaps should be located as specified in the service literature.

Pistons and cylinder walls should be lubricated with clean engine oil prior to installing the pistons. A ring compressor should be used to compress the rings to ease piston installation. During the installation procedure, care should be taken to ensure that the crankshaft journal is not scratched.

After the assembly is installed, the rod bearing-to-crankshaft journal clearance can be checked using Plastigauge.

14. Inspect and/or replace crankshaft vibration damper.

Inspect the harmonic balancer for signs of wear. Also, inspect the rubber mounting for indications of twisting and deterioration. If a balancer has slipped, rotated, or leaks, it must be replaced. If wear to the center bore is present in the form of a groove worn in from the front engine seal, a repair sleeve kit is available. Check the condition of the keyway and replace the balancer if it is cracked.

15. Inspect flywheel/flexplate (including ring gear) and mounting surfaces for cracks, wear, and runout; determine needed repairs.

Inspect the crankshaft flange and the flywheel-to-crankshaft mating surface for metal burrs. Remove any metal burrs with fine emery paper. Be sure the threads in the crankshaft flange are in satisfactory condition. Replace the flywheel bolts and retainer (if fitted) if any damage is visible on these components. Install the flywheel, retainer, and bolts, and tighten the bolts following the torque and sequence provided by the engine manufacturer.

Inspect the flywheel for scoring and cracks in the clutch contact area. Minor score marks and ridges may be removed by resurfacing the flywheel. If deep cracks or grooves are present, the flywheel should be replaced.

Mount a dial indicator on the engine block or flywheel housing and position the dial indicator stem against the clutch contact area on the flywheel. Rotate the flywheel to measure the flywheel runout. If runout exceeds specifications, replace the flywheel.

Insert a finger in the inner pilot bearing race and rotate the race. If the bearing feels rough or loose, replace the bearing. Check a pilot bushing to verify that it is not loose. A transmission input shaft may be positioned in the pilot bushing to check for excessive play. If too much play exists, replace the bushing. A special puller may be used to remove the pilot bearing or bushing. The proper driver must be used to install the pilot bearing or bushing. Always verify that the transmission input shaft fits in a new bushing before attempting to install the engine (or transmission).

Inspect the starter ring gear for excessive wear or damage. On manual transmission flywheels, the starter ring gear is often replaceable. Remove the old gear by drilling a hole through the gear at the "root" between two teeth. Then position a cold chisel between the two teeth and strike it with a hammer. Take note of whether the gear has a chamfer on one side before removing it. To install the new gear, first heat it to about 400°F (205°C) in an oven. Then slip it over the flywheel body and allow it to cool.

D. Lubrication and Cooling Systems Diagnosis and Repair (3 questions)

1. Diagnose engine lubrication system problems; perform oil pressure tests; determine needed repairs.

When performing oil pressure tests, remove the sending unit from the engine. Use the appropriate adapters and connect a test unit gauge to the engine. Start the engine and observe the pressure at idle. Watch the gauge while the engine warms up and note any pressure loss due to the temperature increase. Increase the engine speed to 2,000 rpm while observing the gauge. Compare the test results with the manufacturer's specifications.

When low oil pressure is evident, first check the oil level. Low oil level will cause the oil pump to aerate and lose volume. If oil level is too high, it may be caused by gasoline entering the crankcase from a leaking fuel injector. If the level and the condition are not in question, use a mechanical gauge to check oil pressure.

No oil pressure indicates that there is a problem with the oil pump drive mechanism or the oil pump itself. Other possible causes of insufficient oil pressure include:

- Restricted oil pump pick-up
- Leaking oil gallery plugs
- Improperly sealed pick-up tube
- Lower-than-specified oil level
- Improper oil viscosity
- Sticking or weak oil pressure relief valve
- Worn engine and camshaft bearings, or oil pump

2. Disassemble and inspect oil pump (includes gears, rotors, housing, and pick-up assembly); measure oil pump clearance; inspect pressure relief devices and pump drive; determine needed repairs.

The oil pump is often replaced when the engine is rebuilt. If reused, it must first pass a visual inspection where clearances are checked. If the oil pump fails inspection, some manufacturers provide a rebuild kit. Proper inspection requires the oil pump to be cleaned and disassembled. Most often, the pump is replaced rather than rebuilt because of the low cost of a new pump. Broken oil pump drives can be the cause of no oil pressure. Occasionally, the oil pump will ingest foreign particles, which are too large to pass through the gear assembly. This condition causes the pump to lock up, which will cause the drive shaft to break. Check for shaft twist on inspection.

3. Inspect, clean, test, reinstall/replace oil cooler, by-pass valve, lines and hoses.

Oil coolers can fail due to leaks or restrictions. Hoses must be well located to eliminate wear from vibrations and/or chafing effects. Hoses deteriorate internally and may cause restriction of fluid flow. Test the efficiency of the cooler and hoses; use an infrared heat detector to determine the temperature difference from inlet hose to outlet hose. Perform the test under load conditions after returning to the shop. If the engine has suffered a major bearing failure, it is recommended practice to replace the engine oil cooler. The metal particles can lodge in the cooler's internal passages and be impossible to remove during the cleaning process. If these particles later become dislodged, the result can be catastrophic engine damage. The oil cooler by-pass valve should be moved without binding with only the force of the return spring acting against it. A stuck valve can prevent the oil from passing through the cooler or, conversely, allow the oil to always by-pass the cooler.

4. Change engine oil and filter(s); add proper type, viscosity, and rating of oil.

When changing the oil and filter, attention to detail is important. It cannot be overemphasized to use the correct viscosity engine oil. Some manufacturers now require very specific engine oils such as 0w-20, or specialized synthetics manufactured to such exacting tolerances that there will be only one or two oils marketed to meet those specifications. In some instances, the oil will only be available from the dealer. Using the incorrect oil can void the warranty and result in engine damage. Engine oil should only be poured from clean, sealed containers. If the oil is stored in bulk containers and transport cans are used to move the oil to the engine, extra care must be used to ensure that these items are kept clean.

Oil filters have come in the spin-on replaceable canister style for many years. In this style, the filter is permanently installed in a canister, and the technician simply screws the old one off, cleans the seal contact area on the engine, wipes clean oil on the new seal, and then reinstalls the filter and tightens ¾ of a turn after the seal contacts the base. Many manufacturers have now moved to the replaceable cartridge-type filter similar to what was used prior to 1965 on most engines. In this system, the filter element is removable from the housing. The only items discarded during the oil change are the filter cartridge itself and the sealing o-rings. This style of filter produces less waste; however, it does complicate oil filter servicing. Since the new filter itself is exposed to dirt during handling, care must be taken to ensure the new filter element is not contaminated. The new o-rings must be installed in the appropriate location and lubricated. The filter will have a torque specification that must be observed.

5. Inspect and reinstall/replace pulleys, tensioners, and drive belts; adjust drive belts and check alignment.

Because the friction surfaces are the sides of a V-belt, the belt must be replaced if the sides are worn and the belt is contacting the bottom of the pulley. Belt tension may be checked with the engine shut off and a tension gauge placed over the belt at the center of the belt span. A loose or worn belt may cause a squealing noise when the engine is accelerated. Measuring belt deflection with the engine shut off can also check belt tension. Use your thumb to depress the belt at the center of the belt span. If the belt tension is correct, the belt should have ½ inch deflection per foot of belt span.

Ribbed V-belts usually have a spring-loaded belt tensioner, with a belt wear indicator scale on the tensioner housing. If a power steering pump belt requires tightening, always pry on the pump ear, not on the housing.

6. Diagnose engine cooling system temperature and pressure problems; determine needed repairs.

A pressure tester may be connected to the radiator filler neck to check for cooling system leaks. Operate the tester pump and apply 15 psi (or recommended pressure) to the cooling system. Inspect the cooling system for external leaks with the system pressurized. If gauge pressure drops more than specified by the vehicle manufacturer, the cooling system has a leak. If there are no visible external leaks, check the front floor mat for coolant dripping out of the heater core. When there are no external leaks, check the engine for internal leaks.

The radiator pressure cap may be tested with the pressure tester. When the tester pump is operated, the cap should hold the rated pressure. Always relieve the pressure before removing the tester. A large increase in pressure when the engine is started usually indicates head gasket failure.

Restricted cooling system passages can result in engine overheating. Restricted passages in a radiator can be determined by using an infrared temperature gun and looking for cool spots on the radiator. Cool spots indicate restricted flow.

7. Inspect, test, and replace thermostat, coolant by-pass, and thermostat housing, seals, hoses, and fittings.

If a customer brings in a vehicle with an overheating problem, it is possible that the thermostat is not opening. An engine that fails to reach operating temperature could have a thermostat that is stuck in the open position. Check the temperature rating of the thermostat that is to be replaced, and confirm that it is the proper one for the engine application. Visually inspect the thermostat for rust and other contamination. Make sure that the thermostat was installed properly.

To test the thermostat, submerge it into a container of water. Use a thermometer to determine the temperature when the thermostat opens. Heat the water while observing the thermostat. At the rated opening temperature of the thermostat, it should begin to open.

The coolant by-pass and thermostat housing should be inspected for cracks and restrictions. Hoses are typically replaced during thermostat service; if not, they should be inspected for swelling, hardness, abrasion wear, and internal cracking. Fittings should be checked for damage and corrosion. Corrosion can be a sign of improperly treated coolant.

8. Inspect and test coolant; drain, flush, and refill cooling system with recommended coolant; bleed air from cooling system as required.

Some truck manufacturers recommend premixing the coolant and then filling the system. There are two methods used to check the freeze protection level of the coolant. One is with a hydrometer, which compares the coolant by weight to that of water. The other method, which is considered more accurate and is becoming more highly recommended by engine manufacturers, is the use of a refractometer. It is important that the coolant be the correct mixture not only for freeze protection, but also for other benefits such as heat transferability, corrosion protection, and rust inhibiting. The pH level of the coolant can be tested with litmus paper strips. There are several types of coolant being used, and they are usually incompatible, so make sure any coolant added to the system is correct. Consult engine compartment labels or service manuals to be sure which type is used.

Flushing of the cooling system is accomplished by using pressurized water through the cooling system in a reverse direction of normal coolant flow. A special flushing gun mixes low-pressure air with tap water. Reverse flushing causes the deposits to dislodge from the various components. They can then be removed from the system. The engine block and radiator should be flushed separately.

To flush the radiator, drain the system and disconnect the upper and lower hoses. Attach a long hose to the upper hose outlet to deflect the water. Disconnect and plug the heater hoses that are attached to the radiator. Fit the flush gun to the lower hose opening; this causes the radiator to be flushed in the reverse direction. Fill the radiator with water and turn on the gun in short bursts. Continue flushing until the water exiting the radiator is clean.

The cooling system is filled with a 50/50 mix of antifreeze to water. Before filling the cooling system, make sure that the drain plug and all hose clamps are tight. The preferred method of filling the cooling system is to premix the coolant at a 50/50 mixture, then install it in the engine. Be sure to identify which type of coolant is being used and follow the correct procedure. Many late-model vehicles require bleeding when refilling to eliminate trapped air pockets. To do this, you must open the cooling system by removing a hose or fitting on the engine side of the thermostat and then filling the system until liquid appears, at which point you replace the hose or fitting. Continue to fill the cooling system as needed as the engine warms up. Another approved method to install coolant in the engine is to use a vacuum. This method involves using a special tool to draw a vacuum on the empty cooling system and then allow the vacuum to draw the coolant into the engine. This method is most likely to result in a complete fill without air pockets in the coolant.

9. Inspect and replace water pump, housing, and hoses.

A water pump is replaced when the bearing has failed, or more commonly when the front lip seal has failed and caused a coolant leak. Most water pumps can be removed after disconnecting the lower or upper radiator hoses at the pump. Remove any by-pass or heater hoses that are attached to the pump. Remove the bolts attaching the pump to the front cover. When removing the bolts, take time to keep them in order since they are usually different lengths. The water pump might need a tap with a hammer to separate it from the cover. Remove all traces of gasket material from the front cover and install the water pump using a new gasket. Some water pumps now require removal of the timing belt or the timing chain, which drive the water pump.

10. Inspect and replace radiator, pressure cap, expansion tank, and coolant recovery system.

Inspect the radiator for insects and debris in the fins and for bent and damaged fins. Look in the neck and the upper and lower hose nipples to examine for accumulation of deposits clogging the tubes. Check for evidence of seepage or leaks where the core is attached to the tanks. If the radiator requires service, use the steps below to remove it.

1. Disconnect the negative battery cable.
2. Drain the cooling system.
3. Loosen the hose clamps and disconnect the hoses from the radiator.
4. If equipped, disconnect the transmission cooler lines.
5. If equipped, disconnect the electric fan connector.
6. Remove the bolts attaching the shroud to the radiator.
7. If the air conditioning condenser cannot be disconnected from the radiator, the air conditioning system must be recovered.
8. Remove the upper radiator cross mount.
9. Remove the fan module or the shroud.
10. Remove the radiator.

The coolant recovery system should be checked to ensure that the hoses are tight and in good condition and that the expansion tank is undamaged. Many expansion tanks are plastic and are subject to cracking, so look for leaks along the seams and around the hose nipple.

11. Inspect, test, and repair/replace fan (both electrical and mechanical), fan clutch, fan shroud, air dams, and cooling fan electrical circuits.

The cooling fan can be driven from an accessory belt or may be electrically operated. Regardless of how it is powered, the fan blades must be inspected for stress cracks. Since the fan blades are balanced to prevent vibration, if any are damaged, the fan must be replaced. Belt-driven fans use either flex fans or a viscous fan clutch. Flex fans should be inspected for stress cracks, while viscous fan clutches should be inspected for indications of leakage. Also, inspect the thermostatic spring for free movement and accumulations of dirt and debris. If any free movement exists, replace the clutch.

Trucks may be equipped with one or more electric cooling fans. If there are two fans, some are designed so that one will operate first and the other one will be used if the first fan is unable to bring coolant temperature into normal operating range. Other systems will operate the fans at the same time. A relay is usually needed to transfer high current to the fan motor. The relay may be activated upon command from a sensor or from a computer engine management system. Consult a wiring diagram to locate the power supply, ground points, and control circuit inputs and outputs to understand how the system functions. Many relays are groundside switched with the ECM supplying the ground. Other systems like the HVAC (heating, ventilation and air conditioning) may also interact to regulate fan operation.

The fan shroud should be complete and wrap all the way around the fan. A missing fan shroud can cause overheating. Air dams are often missing. A missing air dam can cause overheating.

12. Verify proper operation of engine related gauges and warning indicators; determine needed repairs.

Operating temperature can be indicted to the driver either by a gauge or a light. The light typically indicates only an overheat condition, while the gauge indicates approximate operating temperature. The light will have a temperature-operated switch to control it. The gauge will have a thermistor controlling it. To check light operation, make sure the light comes on during bulb check, then remove the wire from the sensor, ground it, and see if the light comes on. If it does, it is operating properly. To test the gauge, remove the wire at the sending unit and install the appropriate resistor, as indicated by the service literature, between the wire and ground. Check to see that the gauge reading indicates the appropriate temperature. Some manufacturers will use the information from the sensors that send information to the engine ECM to operate the dash gauges. In this type of arrangement, the dash is on the vehicle data bus and the information is transmitted from the engine ECM to the instrument panel control module. This is sometimes referred to as a serial dash or a two-wire instrument panel, because all the information for the dash is routed through the two wires of the data bus.

E. Ignition System Diagnosis and Repair (6 questions)

1. Diagnose ignition system related problems such as no-starting, hard-starting, engine misfire, poor driveability, spark knock, power loss, reduced fuel economy, and emissions problems; determine root cause; determine needed repairs.

The above-listed conditions should be confirmed and then diagnosed as follows:

1. Is there sufficient battery voltage to provide proper primary and secondary voltage?
2. Is the proper voltage available at the requisite points?
3. Is battery voltage sufficient to provide sufficient engine RPM, as some engines do not deliver spark or fuel until a set engine speed?
4. Are all components in good working order?
5. Have specifications for timing, fuel delivery, air restrictions, etc., been checked beforehand?
6. Have all ground connections been confirmed?

When all of these performance checks have been performed, compare the results to specifications.

2. Interpret ignition system related diagnostic trouble codes (DTCs); determine needed repairs.

Perform testing to confirm both active and historic/inactive DTCs. Do so by means of self-diagnostic (check engine light) and/or diagnostic scan tool. Because trucks with gasoline engines will be found predominantly in the smaller class 3 and 4 categories, those manufactured after 1996 are required to be OBD-II compliant. Some of the features of OBD-II are standardization of the diagnostic connector, its location, and its terms and codes. OBD-II systems run tests called *monitors* and log the results in diagnostic memory. A wide variety of information is available to the technician. Familiarize yourself with retrieving and utilizing information from OBD-II compliant trucks.

3. Inspect, test, repair, or replace ignition primary circuit wiring and components.

The primary circuit of the ignition system is the circuit components other than spark plugs, spark plug wires, and the ignition coil secondary windings.

To perform primary circuit testing, it is advisable to consult the manufacturer's recommended procedures, as damage to electronic/electrical systems can result from incorrectly performed tests. Note that most testing on late-model vehicles can be easily performed with a hand-held scan tool and DMM. On older systems, voltage spikes are not as critical and other test instruments can be used.

Testing should always start with a battery voltage condition test then proceed to component and resistance tests. Replace components that do not conform to the manufacturer's specifications.

To check ignition coil resistance, begin with the key off and the wires disconnected from the coil. Resistance specifications vary among different manufacturers, so be sure to follow their specifications and procedures. The following examples are given only to describe how a typical test would be performed. To check the primary coil winding, connect the test leads to the primary coil terminals. Most coils have very low primary winding resistance, typically about 0.5 to 2.0 ohms. Then check the secondary winding resistance with the ohmmeter and look for very high resistance, typically between about 8,000 and 20,000 ohms. Some coil secondary resistance tests are measured between one of the primary terminals and the secondary terminal, while others are tested between the secondary terminal and the coil frame, so be sure to follow the correct manufacturer's procedures.

4. Inspect, test, repair, or replace ignition system secondary circuit wiring and components.

Secondary circuit components include the spark plugs, the spark plug wires, and the ignition coil secondary windings.

Ignition system secondary circuit wiring should be inspected visibly for chafing against metal brackets or exhaust manifolds, which could cause arcing as well as improper routing. Misrouted wires can increase the possibility of cylinder crossfire, which can cause extreme engine damage. Wires and spark plug boots should also be checked for coolant or oil-soaked conditions and replaced if these conditions exist. Spark plug wire terminals should also be checked for signs of corrosion or arcing at both ends of the wire. Wires can be checked for proper resistance with an ohmmeter; specifications are listed in the service manual and are usually expressed in ohms per foot.

Distributor caps and rotors should be inspected for burned terminals and cracks as well as secondary voltage arcing damage. An ignition oscilloscope can be used to check for leakage in secondary wiring. Spraying a fine mist of water on the wires to simulate damp conditions that cause many secondary ignition problems should be done. Worn spark plug gaps or excessive distributor cap-to-rotor air gaps are easily tested with an ignition scope.

Secondary circuit components are typically replaced, not repaired. During replacement, ensure all mounting hardware is properly installed. All electrical connections should have electrical grease applied to ensure good corrosion connections. During spark plug removal, some engines' plugs are prone to break off in the head. Consult one of the TSBs addressing repair techniques when this happens. Spark plugs should be torqued to the correct specification when reinstalled.

■ 5. Inspect, test, and replace ignition coil(s).

Perform a visual inspection as part of the diagnostic procedure to help determine if the ignition coil is defective. As this is the high voltage source, it is dependent upon a proper low voltage source to complete its task. Determine proper primary voltage. Determine if moisture is collecting on the coil assembly and providing a path for voltage loss. Test output using a spark tester to determine secondary voltage output. Confirm with manufacturer's specifications. Replace and retest with new components as necessary; some units can be replaced on an individual basis.

■ 6. Inspect, test, and replace ignition system sensors; adjust as necessary.

There are several different styles of primary triggering devices used on ignition systems. Magnetic pulse generators, Hall-effect switches and optical pick-ups are the most common.

When testing magnetic pick-ups, an ohmmeter can be used to test for proper resistance of the coil and for a grounded or open coil by connecting the ohmmeter leads across the coil leads. Connect one meter lead to a pick-up coil lead and the other to ground to test for a grounded coil. Other tests that can be performed on magnetic pick-ups include testing AC voltage output with a DMM and signal waveform measurements with a scope. Consult the service manual for specifications for these tests.

Hall-effect switches and optical pick-ups require a voltage feed and ground to operate properly. Once proper feed voltage and a good ground are verified, the signal line can be tested. These pick-ups are often supplied a reference voltage from the ignition control module or powertrain control computer, usually between 5 and 10 volts. A labscope is the best method for testing the signal from these types of pick-ups.

Some Hall-effect and optical pick-ups may have four wires. In this case, there will be two signal lines on the labscope trace; often this will be a low and a high data rate signal. Some optical pick-ups generate a square wave signal output for every degree of camshaft rotation, so the signal created will be at a very high frequency. While a DMM that measures frequency can determine if a signal is being generated, a labscope provides a much better picture of signal integrity. The square wave signal from a Hall-effect or optical pick-up should reach 90 percent of reference voltage when the signal is high and pull down to within 10 percent of ground potential when the signal is low. On a system using an adjustable magnetic pulse generating sensor, the air gap must be set. This is done using a brass or metallic feeler gauge and is usually set at 0.008 in. (0.203 mm). A steel feeler gauge will cause a false drag due to the permanent magnet in the sensor. Some magnetic pulse generating sensors used as crank and cam sensors require the use of a paper spacer to position the sensor with a predetermined gap.

Ignition system sensors such as cam and crank position sensors are usually installed in a bore. This bore should be clean and corrosion free so the sensor can be installed fully seated in the bore. The mounting fasteners must be torqued correctly.

■ 7. Inspect, test, and/or replace ignition control module (ICM)/powertrain/engine control module (PCM/ECM); reprogram as needed.

Many ignition module testers are available from vehicle and test equipment manufacturers. These testers check the module's capability of switching the primary ignition circuit on and off. On some testers, a green light is illuminated if the module is satisfactory, and the light remains off when the module is defective. Always follow the manufacturer's recommended procedure. If the module tests satisfactorily, then you should perform circuit tests to confirm that the wiring of the circuit is serviceable and the proper signals are going to the correct designations.

The ignition module removal and replacement procedure varies, depending on the ignition system. Always follow the manufacturer's recommended replacement procedure. Some ignition modules require the use of dielectric silicone grease for heat dissipation through the mounting surface. Clean the mounting surface and apply a light coat of silicone grease to the module prior to installation. If silicone is not used, heat generated by the module's transistor will not be properly dissipated, and early failure of the module may occur.

Replacement of a PCM should be done only after power and ground circuits to the PCM have been tested and all PCM outputs have been checked for proper current draw to prevent repeated failure. You must be careful to ground yourself to dissipate any static electrical charges from your body prior to handling the replacement PCM. This will prevent damage to sensitive circuit boards from any static electrical discharge. The PCM must be handled with care when replacing to ensure no static discharge into the computer. Most modern-day computers require flash programming to upload new operating parameters for smooth operation. A battery maintainer should be installed to ensure no interruption of battery power when you reprogram a PCM.

F. Fuel, Air Induction, and Exhaust Systems Diagnosis and Repair (6 questions)

1. Diagnose fuel system related problems such as no-starting, hard-starting, poor driveability, incorrect idle speed, poor idle, flooding, hesitation, surging, engine misfire, power loss, stalling, reduced fuel economy, and emissions problems; determine root cause; determine needed repairs.

When diagnosing no-start or hard-starting problems, poor driveability, flooding, hesitation, surging, misfire, power loss, stalling, and emissions problems, always begin by checking for related DTCs, then check for the essentials of combustion. There must be correctly timed spark, correct fuel/air mixture, and adequate compression.

A few quick checks for the more obvious problems include:

- Check for the presence of fuel, and also correct fuel in the tank.
- Check for the presence of fuel pressure at the fuel rail of fuel-injected engines with a pressure gauge.
- If fuel pressure is adequate, verify whether the PCM is controlling the injectors by testing for a signal at the injector with a noid light.
- Further diagnosis of the fuel system may be accomplished using a scan tool.
- A spark tester can be used as a quick check of the ignition system. Both compression testers and vacuum gauges can be used to check the mechanical integrity of the engine.

Sometimes the effect of excess exhaust backpressure is overlooked. A plugged catalytic converter sometimes results in a no-start or hard-starting problem. To check for this, a backpressure gauge is installed in place of the oxygen sensor. If there is more backpressure than 2 psi, locate and repair the restriction. Poor fuel economy with fuel-injected engines can be traced to leaking injectors, defective fuel pressure regulators, and the computer receiving faulty information from its sensors. Using a scan tool displaying a data stream allows you to compare sensor values with just one hook-up.

Idle quality problems such as low or high idle speeds can usually be attributed to vacuum leaks at hoses and gaskets.

2. Interpret fuel or induction system related diagnostic trouble codes (DTCs); analyze fuel trim and other scan tool data; determine needed repairs.

Verify active and/or historic (inactive) DTCs by means of self-diagnostic (check engine light) or by diagnostic scan tools. Often, DTCs point to a system, not a component, and you must consider what the computer sees in order to set a DTC. An example would be a lean exhaust DTC. If the oxygen sensor fails, it no longer produces an output voltage. The computer sees the low signal voltage and interprets this as a lean exhaust signal. The computer will increase fuel delivery in an attempt to generate a high voltage signal from the oxygen sensor. The engine can be running very rich and an exhaust gas analysis will confirm this, yet the computer DTC is set for a lean exhaust signal because of the failed oxygen sensor. When you diagnose fuel trims, zero percent means no fuel is being added or subtracted to keep the air/fuel ratio at 14.7:1. As fuel is added, the fuel trim number will increase; as fuel is decreased, the fuel trim number will increase with a minus sign in front of the number. Fuel trim is okay until the number gets into the teens, and then a closer inspection is needed to determine the cause of the increased or decreased fuel trim.

3. Inspect fuel tank, filler neck, and fuel cap; inspect and replace fuel lines, fittings, and hoses; check fuel for contaminants and quality.

The fuel tank should be inspected for leaks; road damage; corrosion; rust; loose, damaged, or defective seams; loose mounting bolts; and damaged mounting straps. Leaks in the fuel tank, lines, or filter may cause gasoline odor in and around the vehicle, especially during low speed driving and idling. In most cases, the fuel tank must be removed for service.

Nylon fuel pipes should be inspected for leaks, nicks, scratches, cuts, kinks, melting, and loose fittings. If these fuel pipes are kinked or damaged in any way, they must be replaced. Nylon fuel pipes provide a certain amount of flexibility and can be formed around gradual curves under the vehicle. Do not force a nylon fuel pipe into a sharp bend, because this action may kink the pipe and restrict the flow of fuel.

Obtain a sample of the fuel and examine for dirt or other contaminants. Use a commercially available alcohol tester to determine the percentage of alcohol present in the fuel. Generally, fuel injection systems will tolerate only a limited amount of alcohol in the fuel. Consult a workshop manual for details.

The gas cap and fuel tank filler neck must seal correctly. A leak will set a DTC for the evaporative emissions control (EVAP) system. These leaks are normally found using a smoke machine.

▨ 4. Inspect, test, and replace fuel pump(s) and/or fuel pump assembly; inspect, service, and replace fuel filters.

When testing fuel pumps, pressure and volume tests apply. Fuel must be available to the engine at the correct pressure and adequate volume for proper operation. It is possible to have the correct pressure with little or no volume. Fuel volume is the amount of fuel delivered over a specified period of time. The correct amount is specified by the manufacturer and is generally about 1 pint (0.47 liters) in 30 seconds. Caution should be exercised when performing volume testing, because fuel is discharged into an open container, which creates a risk of fire.

Electric fuel pumps can be tested for proper pressure by connecting a fuel pressure gauge to the Schrader valve test port on the fuel rail. If the vehicle does not have a test port on the fuel rail, the gauge will need to be teed into the system using the correct adapters.

Low pressure can result from a restricted fuel line or filter, a defective pump, or a lack of good voltage supply and ground connections in the pump's electrical circuit. If fuel pressure is higher than specified, a bad fuel pressure regulator or a restricted fuel return line could be the cause. Most fuel injection systems should hold pressure after the engine is turned off. A bad fuel pump outlet check valve, a bad fuel pressure regulator, or a leaking fuel injector can cause a pressure leak down. Follow the manufacturer's test procedures to isolate the failed component.

Fuel filters should be replaced at the manufacturer's specified intervals, or more often, if the fuel quality in the area is poor or if the vehicle has had fuel contamination problems.

If a fuel pump is being replaced, the fuel pump relay should be inspected. The failing fuel pump likely required an abnormally high current draw, which may have damaged the contacts in the relay. Some technicians choose to replace the relay at the same time the fuel pump is replaced.

▨ 5. Inspect and test electric fuel pump control circuits and components; determine needed repairs.

Fuel-injected engines using electric fuel pumps use a relay to transfer high current from the battery to the pump. The computer usually closes the relay and runs the pump for about two seconds when the key is turned on to pressurize the system for easy starting. When the engine receives an RPM signal during cranking, it reactivates the relay, and the pump will run continuously. On some vehicles, in case of a relay failure, there is a redundant power supply to the pump through the oil pressure switch. When oil pressure reaches a predetermined level, the switch closes and power is supplied to the pump. If the relay fails, the engine may still start, but cranking time will be longer because of the time needed to build sufficient oil pressure.

Most electric fuel pumps are located inside the fuel tank, and it is necessary to remove the tank to get access to the pump. Test for a faulty pump by supplying power directly to the pump and observing pump operation. When checking for voltage supply to the pump, remember that disconnecting the connector at the fuel tank and checking open circuit voltage may not tell the whole story. It is better to check the circuit under load to determine if voltage drops exist. An amperage test can be performed at the fuel pump relay to determine if the pump is requiring too much amperage. A pump which has had a high current draw over a long period of time may have damaged the fuel pump relay's contacts. In a scenario such as this, the technician may replace the fuel pump and return the truck to the customer only to have it return a few weeks later with a failed fuel pump relay. In this case, it is most likely that the original failed fuel pump caused damaged to the relay.

6. Inspect, test, and repair or replace fuel pressure regulation system and components of fuel injection systems; perform fuel pressure/volume test.

The fuel pressure regulator should be tested for leakage through the diaphragm by removing the vacuum hose with the engine running to see if fuel drips from the vacuum nipple. Fuel pressure should reach the pressure specified by the manufacturer with the vacuum line disconnected and lower by about 10 psi (68.9 kPa) when the vacuum line is plugged back onto the regulator. Excessive fuel pressure indicates a stuck closed regulator or restricted fuel return line. Low pressure can be caused by a weak fuel pump, restricted fuel filter, or stuck open regulator. If the pressure increases to normal when the return line is restricted, the regulator should be replaced.

If fuel pressure drops when the engine is turned off, a leak is indicated in the fuel system. Alternately clamping off the fuel feed and return lines will isolate the location of the leak. If the pressure drop stops when the return line is pinched, the fuel pressure regulator is leaking and should be replaced. If the pressure still drops, a leak through the fuel pump or injectors is the problem. Clamping the fuel feed line will eliminate the fuel pump as the source of the pressure drop if the system holds pressure, and then the injectors will need service. Do not clamp plastic fuel line, as this will cause permanent damage. Rubber test lines need to be installed to perform this test if the vehicle has plastic fuel lines.

Fuel system components are typically replaced. Ensure that replacement components are the correct parts. Many items are physically the same but internal pressure settings and orifice sizes may be different. The correct part number is essential.

7. Inspect, remove, service or replace throttle body assembly and controls, including electronic throttle actuator control (TAC) systems; make related adjustments.

To remove the throttle body, first remove the air intake assembly connected to the throttle body. Disconnect the throttle linkage and, if equipped, the transmission kickdown cable. Disconnect the throttle position sensor (TPS) and remove the throttle body. Remove the old gasket, since a new one will be installed during installation. Place a rag in the intake manifold opening to ensure that no foreign objects enter the engine. Some throttle bodies have a special coating that prevents carbon buildup on the surface of the throttle plate. If the throttle body has a special coating, no chemical cleaners can be used to clean the throttle body.

Some manufacturers allow cleaning the throttle body assembly; others do not. Always follow manufacturer's recommendations. When a replacement throttle body or electronic throttle actuator is installed, the ECM must go through a relearn process, which will involve driving the vehicle so the ECM can learn to control the engine idle.

8. Inspect, test, clean, and replace fuel injectors and fuel rails.

Fuel injectors can be tested on the truck by performing an injector balance test. While monitoring fuel pressure, each injector is fired one at a time with a special tool designed to open the injector an exact amount of time. When the injector is triggered, the fuel pressure in the rail will drop. No pressure drop indicates a plugged injector or open coil. An injector with a low-pressure drop indicates a dirty or restricted injector. An injector with too great a pressure drop indicates a leaking or rich injector. A pressure difference greater than 1.5 psi (10.3 kPa) above or below the average is considered a problem that requires service. Fuel injectors should also be tested with a labscope to observe their electrical integrity. Both voltage and current waveforms can be observed with a labscope.

Tool manufacturers market a variety of fuel injector cleaning equipment. A solution of fuel injector cleaner is mixed with unleaded gasoline. Shop air pressure provides system operating pressure. The vehicle's fuel pump must be disabled to prevent fuel from being forced into the fuel rail. The fuel return line should be plugged to keep the fuel injector cleaner solution from entering the fuel tank. After the fuel injectors have been cleaned, the adaptive memory will need to be reset.

If the injectors have been removed for cleaning, the spray pattern should be checked. An even cone-shaped pattern without any dripping discharge should be present. If the proper spray pattern cannot be achieved, the injector should be replaced.

Fuel rails should be inspected for leaks, damage and loose mounts. Leaking fuel rails have been the subject of many recalls. TSBs should always be checked when diagnosing a fuel leak. Fuel rails that fail inspection are usually replaced, not repaired.

9. Inspect, service, and repair or replace air filtration system components.

Perform visual inspection on all air intake and/or air preheat systems. Driveability and or emission-related problems may be the result of inoperative or malfunctioning air inlet/preheat systems. Perform testing as per manufacturer's recommendations to confirm proper operation and unrestricted airflow. Air cleaner restriction indicators are widely used throughout the industry. Although they are not noted for their accuracy, they do provide an indication of when filter service is required.

If a vehicle is operated continually in dusty conditions, air filter replacement may be necessary at more frequent intervals. A damaged air filter can cause increased wear on cylinder walls, pistons, and piston rings. If the air filter is restricted with dirt, it restricts the flow of air into the intake manifold, and this increases fuel consumption. Many trucks are equipped with an air filter restriction gauge. The restriction gauge should be checked during vehicle service. If the indicator shows an excess of 20 inches of water, most manufacturers recommend replacing the filter.

The air filter housing should be checked closely for leaks. These housings are plastic, and often the high underhood temperatures can cause warping of the components. Replacement, not repair, is the normal practice.

10. Inspect air induction system, intake manifold, and gaskets for air/vacuum leaks.

The air induction system must seal tightly. Dirt entering the system will result in an engine that is quickly worn out. If the oil analysis is used in the maintenance program, a high silicon content in the oil indicates dirt is entering the engine, possibly through the air intake system. The air intake system is usually checked for leaks using a smoke machine.

11. Remove, clean, inspect, test, and repair or replace fuel system vacuum and electrical components and connections.

Fuel system vacuum and electrical components include the fuel pressure regulator and any vacuum controls, if used, vacuum-operated throttle positioner, fuel pump relay, inertia switch, two-speed fuel pump resistor, and electronic fuel pump power modules. A visual inspection will uncover damaged vacuum lines, and proper routing can be checked against the underhood emissions label. All electrical connections should be visually checked for terminal seating as well as damaged, chafed, or corroded wires or connections. Basic electrical testing with a test light and DMM can determine problems with fuel system electrical components. The service manual should be consulted to identify which components are used and what, if any, special test procedures may be required.

12. Inspect, service, and replace exhaust manifold, gaskets, exhaust pipes, oxygen sensors, mufflers, catalytic converters, resonators, tailpipes, and heat shields.

The exhaust can affect the driveability and emissions of a vehicle, especially one that is computer controlled. Should a leak occur at any area located upstream of the oxygen sensor, the result would be incorrect data signaled by the oxygen sensor and a resultant change in the fuel delivery logic by the ECM.

Incorrect backpressure will affect fuel delivery, driveability, performance, and emissions. One of the easiest test devices to determine restricted exhaust is the vacuum gauge. With the engine idling and the vacuum gauge reading stabilized, if the reading decreases, there may be an exhaust restriction. Open the throttle and then allow it to snap back to the idle position. The vacuum gauge should jump above normal and then quickly return to normal. If it does not, there is a restriction in the exhaust.

Remove the exhaust pipe bolts at the manifold flange, and disconnect any other components in the manifold, such as the oxygen sensor. Remove the bolts retaining the manifold to the cylinder head and lift the manifold from the engine compartment. Remove the manifold heat shield. Thoroughly clean the manifold and cylinder head mating surfaces. Measure the exhaust manifold surface for warping with a straightedge and feeler gauge in three locations on the manifold surface. Examine the manifold carefully for any cracks or broken flanges.

Follow the exhaust system from manifold to tailpipe end. Ensure that all hangers are present and installed correctly. The exhaust system is designed to be suspended from these hangers; loosen joints and realign if any of the hangers are in tension. Examine all pipes, mufflers, and resonators to ensure that they are securely connected and gas tight.

Catalytic converters can be checked for efficiency by comparing the upstream and downstream exhaust gas oxygen sensor values using a scan tool. An alternate method is to measure the inlet and outlet temperature of the converter. Replacement converters must be of the same style as the original. Usually, the difference between low cost and high cost converters is the quality of the components inside the converter.

13. Test for exhaust system restriction or leaks; determine needed repair.

A restricted exhaust pipe, catalytic converter, or muffler may cause excessive exhaust backpressure. If the exhaust backpressure is excessive, engine power and maximum vehicle speed are reduced. Even a partial restriction will reduce performance and fuel mileage. A low range pressure gauge or compound vacuum gauge can be connected to the exhaust system to measure backpressure directly. Backpressure test adapters are available that screw into the oxygen sensor hole and provide a hose nipple for connection of a compound vacuum gauge. Another adapter looks like a rivet that is installed into a hole drilled into the exhaust system front pipe and allows a gauge to be screwed into the inside threads of the adapter. Generally, exhaust backpressure should measure less than 3 psi (20.6 kPa) maximum when engine speed is held around 2500 rpm.

Another method of testing for exhaust restrictions is to measure intake manifold vacuum. Normal vacuum at idle should be between 16 to 21 in. Hg (48.3 to 31 kPa absolute). When the engine is accelerated to 2500 rpm and held, the vacuum reading will drop momentarily and then stabilize equal to or greater than the idle reading. A vacuum reading that drops very low or to zero indicates a restricted exhaust system.

Exhaust leaks can be located using a smoke machine. Exhaust leaks can allow oxygen into the exhaust stream and cause the engine to run rich, resulting in poor fuel economy

G. Emissions Control Systems Diagnosis And Repair (5 questions)

1. Test and diagnose emissions or driveability problems caused by positive crankcase ventilation (PCV) system.

The PCV system requires service at regular intervals to ensure that the system operates properly. Sludge and carbon can plug the PCV valve. Service the PCV valve by inspecting it, making sure that the valve is not clogged and the internal plunger moves freely. Some symptoms of a PCV system that is not functioning properly include the following:

- increased oil consumption
- diluted or dirty oil, or oil that has sludge or is acidic
- excessive blowby, or oil level dipstick that is blown out of its seat
- rough idling or stalling at idle or low engine speeds

2. Inspect, service, and replace positive crankcase ventilation (PCV) filter, valve, tubes, orifice/metering device, and hoses.

Inspect all PCV system hoses for cracks and deterioration. Check the air filter for oil deposits and for oil puddling in the oil cleaner housing. In addition, the system should be inspected for clogs. Clogs are usually caused by not changing the oil at regular intervals and by not maintaining the PCV system. Regular maintenance should include cleaning the PCV valve with a carburetor or fuel injector cleaner. Orifice-type PCV systems should be checked for restriction. Some of these systems will deform from high underhood temperatures and will need to be replaced.

3. Test and diagnose driveability problems caused by the exhaust gas recirculation (EGR) system.

The EGR valve should open once the engine is warm and run above idle or under road load conditions. If the EGR valve remains open at idle and low engine speed, rough idle and stalling can occur as well as engine surging during low-speed driving conditions. When this problem is present, the engine may also hesitate on low-speed acceleration or stall after deceleration or after a cold start. If the EGR valve does not open, engine detonation can occur and emissions levels will increase. EGR system problems can affect emissions levels differently, depending on what the problem is. A stuck open EGR valve will create a density misfire condition in which the air/fuel mixture is diluted with exhaust gas, causing misfire in the combustion chamber. This problem will increase hydrocarbon emissions and raise oxygen levels. An EGR valve that does not open will cause an increase in oxides of nitrogen emissions.

By introducing exhaust gas into the engine during acceleration and cruise conditions, the inert exhaust gas helps reduce peak cylinder combustion temperatures below 2,500°F (1,371.1°C). When temperatures rise above 2,500°F (1,371.1°C) oxygen combines with nitrogen to form oxides of nitrogen (NOx). Oxides of nitrogen are a main contributor to photochemical smog and must be controlled to manageable levels inside the engine, because the three-way catalytic converter is not very efficient at reducing NOx. If the EGR valve does not open, combustion chamber temperatures will rise, and NOx production will increase far beyond what the catalytic converter can control.

The EGR system can be tested using a scan tool. Late-model systems will typically have the ability to command the EGR valve open and closed from the scan tool. A vacuum pump can be used to test for vacuum leaks and proper opening and closing of the solenoid control valves.

4. Interpret exhaust gas recirculation (EGR) related scan tool data and diagnostic trouble codes (DTCs); determine needed repairs.

DTCs for the EGR system will usually be set for one of three reasons: a control circuit fault, a no or low flow condition, or excessive flow or flow when not commanded condition. The DTC should identify the problem area but not necessarily the component at fault. Once the DTC is retrieved, the service manual must be consulted to determine the necessary tests to be performed to isolate the cause of the code.

5. Inspect, test, service, and replace components of the exhaust gas recirculation (EGR) system, including EGR valve, tubing, passages, vacuum/pressure controls, filters, hoses, electrical/electronic sensors, controls, solenoids, and wiring of EGR systems.

The first step in diagnosing any EGR system is to check all of the system's vacuum and electrical connectors. In many systems, the PCM uses inputs from various sensors to operate the EGR valve. Improper EGR operation may be caused by a defect in one or more of the sensors. DTCs should be retrieved and the cause corrected before any further diagnostics are completed.

There are both vacuum-operated and electronic EGR valves in use. Vacuum-operated EGR valves should be tested for proper vacuum supply to the valve. A vacuum pump should be used to apply vacuum to the valve to test the diaphragm and see that the valve opens. Some valves may require that the engine be running and off idle to operate the internal backpressure transducer so that the valve will hold vacuum. A noticeable change in engine speed and idle quality should be observed. This confirms that the EGR passageways are not plugged. No change in engine operation will require removing the valve and cleaning the passages. A scan tool may be necessary to test electronic EGR valves properly. Vacuum bleed filters can be used on some systems and may require periodic replacement if they become clogged.

Often, EGR problems are caused by faulty EGR controls, such as the EGR vacuum regulator (EVR). This regulator can be checked with a scan tool or an ohmmeter. Connect the meter across the terminals of the EVR. An infinite reading indicates there is an open inside the EVR, whereas a low resistance reading means the EVR's coil is shorted internally. The coil should also be checked for shorts to ground. To do this, connect the meter at one of the EVR terminals and the other lead to the case. The reading should be infinite. If there is any measured resistance, the unit is shorted.

Other EGR control components can also be checked on the scan tool or with a DMM. Refer to the appropriate service manual for the exact procedures and the desired specifications.

6. Test and diagnose emissions or driveability problems caused by the secondary air injection or catalytic converter systems.

Catalytic converters have been in use since 1975. They are placed in the exhaust system very close to the engine and are designed to clean the exhaust of excess pollutants through a chemical reaction. There are three basic types of catalytic converters:

1. Two-way converters are used primarily on pre-1980 vehicles and control unburned hydrocarbons (HC) and carbon monoxide (CO).
2. Three-way converters (TWC) control HC, CO, and oxides of nitrogen (NOx). These converters are used on 1981 and later vehicles with computerized engine controls.
3. Three-way plus oxidation converters are used on 1980 and later vehicles with computerized engine controls and air injection.

The most common reasons for converter failure are overheating and contamination from oil burning or leaded fuel. An engine misfire or extremely rich air/fuel mixture can allow unburned fuel to enter the converter, which can cause excessive heat and converter failure.

Various tests are available to determine if the converter is functional. One test is to measure converter inlet and outlet temperatures with a temperature probe to see if the converter lights off (begins operating). A properly operating converter should show a temperature increase at the outlet compared to the inlet. Most manufacturers call for a 10 percent increase in temperature. A temperature increase greater than several hundred degrees could indicate the converter is working too hard due to excessive hydrocarbons present in the exhaust. The engine should be checked for misfiring. You can use an exhaust analyzer to perform a cranking CO_2 test or an oxygen storage test, or take measurements before and after the converter to calculate converter efficiency, also called an *intrusive test*.

Intrusive converter testing should produce efficiency results above 80 percent. The oxygen storage test should show less than a 1.2 to 1.7 percent increase in oxygen readings during snap-throttle acceleration from 2000 rpm if the converter is working properly and storing oxygen. A cranking CO_2 test determines if a pre-heated converter can convert hydrocarbons to carbon dioxide while the engine is cranked with the ignition system disabled. The fuel system must be supplying fuel because that is the source of the hydrocarbons. Tailpipe CO_2 reading should be above 12 percent, and hydrocarbons should stay below 500 parts per million (ppm).

Secondary air injection is used to reduce hydrocarbon (HC) and carbon monoxide (CO) emissions by oxidizing these pollutants in the exhaust manifold or catalytic converter. On some vehicles, outside air is injected into the exhaust manifold or converter by a belt-driven or electric air pump or by a pulse air injection system. The air is routed through hoses and pipes by control valves and one-way check valves during certain engine operating conditions and mixed with exhaust gases as they leave the engine. Check valves installed at the exhaust manifold and catalytic converter air supply pipes prevent exhaust gases from flowing back into the control valves or air pump and damaging these components.

Air injection control valves consist of a diverter valve (used to dump air pump output to the atmosphere during deceleration to prevent backfiring) and switching valves that send air pump output to either the catalytic converter or exhaust manifold, depending on engine operating conditions.

On closed-loop feedback fuel control systems, the air pump output is directed to the exhaust manifold after starting and during warm-up. This allows faster warm-up of the oxygen sensor and catalytic converter. Once closed-loop fuel control is entered, the air pump output is switched to the catalytic converter to provide additional oxygen for the rear bed or oxidation bed of the converter. If a dual bed converter is not used, the air pump output is diverted to the atmosphere. If the air pump fails or the hoses are disconnected, tailpipe emissions will increase. If air is not diverted away from the exhaust during deceleration conditions, exhaust backfiring may result.

Another type of secondary air injection is known as a pulse air injection system. On this system, outside air is drawn into the exhaust manifold by negative pressure pulses, created as the exhaust is pushed out of a cylinder by the piston. This system requires no power from the engine to run a pump, as it does in the belt-driven varieties.

Typically, the secondary air injection system can be operated through the scan tool. When the command is sent to deliver air, if the system is operating correctly, the technician should see an appropriate change in oxygen sensor voltage.

7. Interpret secondary air injection system related scan tool data and diagnostic trouble codes (DTCs); determine needed repairs.

Some vehicles can set DTCs for secondary air injection system problems, such as control circuit fault codes, and airflow switching problems, such as air not being delivered to the exhaust manifold when commanded. Other vehicles may have no computer diagnostic capabilities for the secondary air injection system. Consult the service manual for specific models. Catalytic converter efficiency is monitored on OBD-II compliant vehicles and will set DTCs if the converter becomes degraded. Follow manufacturer-specific test routines if a catalyst DTC is set.

8. Inspect, test, service, and replace mechanical components and electrical/electronically-operated components and circuits of secondary air injection system.

Check all hoses and pipes in the system for looseness and rusted or burned conditions. Burned or melted air injection hoses or valves indicate leaking check valves. Inspect the check valves and replace them if they show signs of leakage. With the engine idling, listen for noises from the pump (if equipped). Check the air pump drive belt; adjust it if loose, or replace it if worn or damaged. Check for adequate airflow from the pump and test for airflow to the exhaust manifold during engine warm-up and for flow to the catalytic converter when the fuel system enters closed loop.

A properly operating air pump should raise tailpipe oxygen readings above two percent and often show levels as high as three to eight percent. This can be verified by watching the scan tool oxygen sensor. Airflow should also divert to the air cleaner or atmosphere when the engine is decelerated rapidly or during high RPM operation. This is the function of the air pump diverter valve. The air injection switching valve controls air flow to the exhaust manifolds or catalytic converter, depending upon engine operating conditions. On some computer-controlled vehicles, the PCM can control the diverter and switching valves according to engine operating conditions.

The computer commands to the air injection system valves can be checked on a scan tool. A scan tool may allow testing on some electric air pumps by means of allowing you to turn the air pump on and off. Check the vehicle service manual to identify system components and determine proper test procedures.

Secondary air injection components require no maintenance other than servicing the drive belt if the system has one. Components are typically replaced when they fail.

9. Inspect catalytic converter. Interpret catalytic converter related diagnostic trouble codes (DTCs); analyze related scan tool data to determine root cause of DTCs; determine needed repairs.

When a catalytic converter starts to fail, it is very common for a DTC to be set for degraded catalytic converter efficiency, During diagnosis of this code, the technician will use the scan tool to compare the oxygen sensor voltages before and after the catalyst. If the catalyst has failed, in addition to changing the converter, the technician will need to do a thorough inspection of the engine to ensure it is running properly and will not damage the new converter. Catalytic converters do eventually need to be replaced. Some root causes of early failure include over-fueling the engine, insufficient airflow through the engine, and excessive engine oil consumption.

If the catalytic converter rattles when tapped with a soft hammer, the internal components are loose and the converter should be replaced. When a catalytic converter is restricted, a significant loss of power and limited top speed will be noticed.

10. Test and diagnose emissions or driveability problems caused by the evaporative emissions control system.

The EVAP system captures and stores vapors from the fuel system in a charcoal canister to be burned once the engine is started. These vapors are purged from the canister by engine vacuum after the engine is started and run off idle. If the engine purges vapors from the charcoal canister during idle, a rich condition and possible rough idle could result. Fuel-saturated charcoal canisters may cause excessively rich air/fuel ratios during acceleration, which may cause state emissions test failures. Leaks in the EVAP system can cause customer complaints of gasoline odors in or around the vehicle.

Some vehicles have an enhanced EVAP system that can monitor purge flow rate as well as determine if the system has fuel vapor leaks. Problems with system purging or leaks will set DTCs on these systems

Failures of the EVAP system are typically leaks. Leaks are found using a smoke machine, which generates smokes and injects it into the system while the technician watches for the smoke trail. Occasionally, solenoids in the system will fail, and these problems are typically diagnosed using a scan tool and a DMM.

11. Interpret evaporative emissions-related scan tool data and diagnostic trouble codes (DTCs); determine needed repairs.

Once DTCs are retrieved, you need to consult the service manual for proper diagnostic procedures and system operation. DTCs will identify whether the problem is with canister fuel vapor purging or system leaks. Because the PCM can identify very small leaks and store a DTC, special test equipment may be needed to diagnose and locate the problem.

12. Inspect, test, and replace canister, lines, hoses, mechanical, and electrical components of the evaporative emissions control system.

A careful visual inspection of the EVAP system should be performed if a customer complaint of gasoline odors is received. A gas analyzer or smoke machine will help identify any leaks. Often, hoses will be damaged or left disconnected after other repairs are done. EVAP system purge and vent solenoids need to be checked for proper resistance and electrical operation, as well as proper mechanical operation.

Most purge solenoids are normally closed and block engine vacuum to the canister when off. Energizing the purge solenoid will allow vacuum through to purge the canister. Vent solenoids are normally open and allow air to pass through the solenoid until the solenoid is energized. The vent solenoid is used to seal the system so the PCM can test for system leaks. Charcoal canisters may be equipped with filters that should be replaced at the manufacturer's specified interval.

The fuel tank cap should be carefully inspected for proper application and sealing if a system leak code is set. This is one of the most common problems for setting EVAP system leak codes. Special testers are available to test the pressure and vacuum valves in the gas cap. Fuel tanks may also be equipped with a rollover valve to prevent fuel from escaping through the EVAP system if an accident causes the vehicle to turn upside down or roll over.

Some EVAP systems may use a tank pressure control valve (TPCV) that controls the flow of vapors to the charcoal canister. If fuel vapor pressure in the fuel tank is below 1.5 in. Hg (5.0 kPa), the valve will be closed and fuel vapors will be stored in the tank. When vapor pressure exceeds the set point of the valve, the vapors are vented to the charcoal canister. The TPCV also provides vacuum relief to protect against vacuum buildup in the fuel tank.

H. Computerized Engine Controls Diagnosis And Repair (8 questions)

1. Research OBD-II system operation to determine the enable criteria for setting and clearing diagnostic trouble codes (DTCs) and malfunction indicator lamp (MIL) operation. Perform appropriate drive cycle to determine system condition and verify repair effectiveness.

Generally, when a fault has occurred that causes emissions levels to increase one and a half times the federal limit, a DTC will set and the MIL will illuminate. Certain conditions, such as a catalyst-damaging fault (a severe misfire, for example), may cause the MIL to flash. To test the operation of the MIL, turn the key on and confirm that the MIL illuminates for approximately two seconds. This verifies that the MIL lamp and circuit are operational. Refer to vehicle-specific information for a detailed description of MIL operation.

A DTC that is set in the PCM indicates that a certain test has run and failed. A diagnostic flow chart for the specific DTC will generally aid you in diagnosing the cause of the DTC. Along with the diagnostic chart should be information about the DTC, such as the conditions necessary to run the test (also known as enabling criteria), what conditions cause the DTC to set, and the conditions necessary for the test to pass and clear the DTC.

After repairs are completed, it will be necessary to drive the vehicle to ensure no DTCs are set. This will involve driving the vehicle through many different operations, such as idle, stop and go and highway driving. After completing the test drive, the technician will need to connect a scan tool to ensure that the appropriate drive cycle monitor has completed.

2. Interpret OBD-II scan tool data stream, diagnostic trouble codes (DTCs), freeze frame data, system monitors, monitor readiness indicators, and trip and drive cycle information.

Retrieving DTCs varies greatly among the many different manufacturers' vehicles. Manual code retrieval on pre-OBD-II vehicles includes some of the following procedures: installing a jumper wire across the proper terminals of a diagnostic connector and counting lamp flashes, cycling the ignition switch on and off three times in a five-second period to signal the PCM to enter diagnostic mode and counting lamp flashes, and turning a switch on the PCM to enter diagnostics and counting the blinking of LEDs in the PCM. Many pre-OBD-II vehicles also support scan tool code retrieval.

OBD-II vehicles require the use of a scan tool to retrieve and clear DTCs. The scan tool will also allow viewing freeze frame data that is stored in the PCM when a DTC is set. A PCM runs certain monitors to test components such as the catalytic converter, oxygen sensor heater, and EGR systems. Certain enabling criteria must be met in order for the PCM to run a monitor. If all enabling criteria are not met, the monitor will not be run. Enabling criteria include certain temperatures, certain speeds, and certain time frames. Each monitor's enabling criteria are different.

3. Read and interpret technical literature (service publications and information including wiring schematics).

Gathering service information begins with understanding and verifying the driver's complaint or the vehicle's symptoms. You should also find out if any recent service work has been done on the vehicle or if any aftermarket electronic accessories have been installed. If custom audio equipment or other electronic equipment has been installed, these accessories may be the source of electromagnetic interference (EMI) that affects the control system. Additionally, a poor ground connection or power source for an accessory can cause problems with a vehicle's original electronic equipment. Learning about the service history and condition of a truck is the act of getting firsthand information about the vehicle.

Service information for the PCM system can come from the same sources as other information about the truck. Selecting the most relevant information can be the key to a fast and accurate diagnosis. Wiring diagrams may be particularly important for control system troubleshooting. Diagnostic flow charts and "trouble trees" can be equally important.

Nearly all manufacturers now provide all service publications and wiring schematics online. This information is available to the dealership technician as well as the independent technician. The independent technician is required to pay a daily, weekly, or monthly fee to access this literature.

4. Diagnose the causes of emissions or driveability problems with stored or active diagnostic trouble codes (DTCs).

Once a fault has been detected by the PCM, it stores a DTC in memory, and if the fault affects exhaust emissions, it will light the MIL. You then retrieve the stored DTC and access a diagnostic flow chart, which leads you through a series of steps to determine the actual problem.

When you diagnose a fault, it is useful to be aware of the specific set of circumstances that caused the control module to set a fault. This information, found in the workshop manual, will allow you to further refine the diagnostics necessary to solve the problem.

An important step in the process of diagnosing computer DTCs is to determine if the code is a history (memory) code or a current code, which means the fault was present at the time the DTC was retrieved. Following a DTC flow chart when a code is a history code can cause misdiagnosis and replacement of unnecessary parts. Most late-model PCMs will report codes as either "current" or "history." On early-model systems that do not differentiate between current and history codes, the code should be cleared and the vehicle driven to see if the code resets. On OBD-II systems, the codes should not be cleared until the vehicle is repaired to prevent the PCM from erasing stored freeze frame data.

5. Diagnose the causes of emissions or driveability problems without diagnostic trouble codes (DTCs).

Vehicles with computerized engine control systems may exhibit many driveability or emissions problems without setting a DTC. A hesitation on acceleration can be experienced from a faulty TPS. A bad spot in the sensor circuit board could be causing the signal voltage to momentarily drop. The computer interprets this as a decrease in throttle position, when actually the vehicle is still accelerating. The computer may never set a DTC based on this type of fault, because the voltage never varies above or below the voltage levels needed to set a TPS code.

A leak in the vacuum hose to a manifold absolute pressure sensor (MAP) may cause a higher-than-normal reading from the MAP sensor. The computer interprets the reading as higher engine load and increases fuel delivery. Higher exhaust emissions will result, but a DTC may not be stored, because the MAP sensor voltage output is still within the normal operational range.

These conditions require careful pinpoint testing to identify the root cause of the problem. An understanding of normal system readings and specifications is needed so you can diagnose problems that do not set DTCs in the computer. Specialty tools such as power graphing multi-meters, labscopes with record modes, and graphing scan tools make the job of finding intermittent circuit or sensor problems much easier.

6. Use a scan tool, digital multi-meter (DMM), or digital storage oscilloscope (DSO) to inspect or test computerized engine control system sensors, actuators, circuits, and powertrain/engine control module (PCM/ECM); determine needed repairs.

In order to control engine operation, the PCM must have a certain set of sensor inputs on which to make decisions. Once these inputs are received, the PCM processes the signals and decides which course of action to take. The PCM then outputs signals to a series of actuators that in turn provide the engine with the things it needs to operate efficiently. In order to troubleshoot a system effectively, you need to confirm that the sensor is transmitting the appropriate signal to the PCM and the PCM is transmitting the proper signal to the actuator.

To properly diagnose vehicles, a technician needs to be familiar with and be able to operate sophisticated diagnostic equipment such as various scan tools and DSOs. The scan tool allows a convenient means of accessing computer sensor data as well as the output commands or status of the engine control system. Many scan tools today allow bi-directional testing, in which you can take direct control of items like idle speed or perform testing such as engine power balance by disabling individual fuel injectors. DMMs and DSOs allow detailed circuit and component testing.

Computer and sensor power feeds, grounds, and signal wires can be tested with a DMM. Sensor and actuator waveform analysis is performed with a DSO. Items such as fuel injectors and oxygen sensors are tested most effectively through waveform analysis. Consult equipment and manufacturer's test procedures to utilize these test instruments to their fullest capabilities

7. Measure and interpret voltage, voltage drop, amperage, and resistance using digital multi-meter (DMM) readings.

Using a DMM, you can evaluate circuit integrity by testing available voltage, voltage drop, amperage, and resistance readings. Available voltage tests confirm if a component is receiving the proper amount of voltage. A voltage drop test is used to locate unwanted circuit resistance when the circuit is energized.

Amperage tests can locate shorted or high-resistance actuators. Resistance measurements can be used to compare components to specifications or for testing wiring harness continuity. Connecting a DMM to measure available voltage is done by making a parallel connection across the circuit. The red lead is connected at the desired test point, and the black lead is connected to the negative battery terminal or engine block. When a voltage drop test is performed, the DMM leads are connected across a component or section of a circuit on the same side of the circuit, either the power or ground side of the circuit.

The reading on the voltmeter indicates the amount of voltage used by the component or section of the circuit between the meter leads. Current must be flowing in the circuit for a voltage drop to occur. A high voltage drop across a conductor or connector indicates excessive resistance. A low voltage drop across a load means resistance is present elsewhere in the circuit. Voltage drop testing across the power and ground sides of a circuit will identify the location of the unwanted resistance.

Inserting the meter in series with the circuit performs amperage testing with a DMM. All DMMs are rated for the amount of current they can measure directly. Care must be taken to prevent connecting the meter to a circuit with higher current draw than the meter can measure. Meters are protected with an internal fuse that will open and require replacement if the meter's current rating is exceeded. Current probes are available that will allow a DMM to make high current measurements. Resistance measurements can be made on components that are removed from the circuit. Whenever resistance tests are performed on circuit wiring, battery power must be removed from the circuit under test. All ohmmeters are self-powered and must not be used on live circuits.

8. Test, remove, inspect, clean, service, and repair or replace power and ground distribution circuits and connections.

The power distribution circuit is the power and ground circuits from the battery, through the ignition switch and fuses, to the individual circuits on the vehicle. Connections must be free of corrosion, as it adds unwanted resistance to current flow. Power and ground circuits are tested by checking resistance and voltage drop. If the circuit is totally inoperable, the resistance check would be most appropriate. However, usually the circuits begins to operate intermittently. This type of failure is often associated with a voltage drop. To test the circuit for voltage drop, current must be flowing through the circuit. The normal maximum voltage drop is 0.5 volts. It is desirable to have less than 0.2 volts.

9. Remove and replace the powertrain/engine control module (PCM/ECM); reprogram as needed.

The PCM must be handled with care when it is replaced to ensure no static discharge into the computer. Locate a good ground on the vehicle and connect a grounding strap from yourself to the vehicle ground before installing the component. Most computers require flash programming to upload new operating parameters for smooth operation. A battery maintainer should be installed to ensure no battery loss when reprogramming a PCM.

An example of flash programming goes as follows.

1. Check the calibration history of the vehicle: Go to the vehicle's website and enter the vehicle identification number (VIN) to learn the vehicle's latest program. If the programming has been updated, the most recent calibration will be listed on the website.

2. Before starting the calibration procedure, make sure the battery is fully charged. Some manufacturers do NOT approve using a normal battery charger; instead they require a battery maintainer that it is manufacturer-approved and delivers very consistent voltage with no fluctuations. On some vehicles, you may have to remove the fuses for the fan relays, fuel pump, or other modules to prevent these devices from turning on during the procedure.

3. On the PC, start the recalibration software program and enter the vehicle application information (year, make, model, etc.).

4. Connect the scan tool to the diagnostic connector on the vehicle (located under the dash near the steering column).

5. Validate the VIN.

6. Choose the operating system, engine, fuel system, speedometer, or transmission.

7. Select "reprogramming."

8. Choose "update bulletin/recalibration number" from the menu.

9. Start the transfer of data. As the software is loaded, a progress bar usually appears. The reprogramming procedure may take from a few minutes to 30 minutes or more depending on the file size, and it can be done with the computer in or out of the vehicle.

10. When the software has finished loading, the message "PROGRAMMING COMPLETE" or a similar message will appear.

11. Turn the ignition OFF, and then disconnect the scan tool. Depending on the application, it may be necessary to run one or more "relearn" procedures before the PCM will function properly.

10. Diagnose driveability and emissions problems resulting from failures of interrelated systems (for example: cruise control, security alarms/theft deterrent, torque controls, traction controls, torque management, A/C, and/or non-OEM installed accessories).

Vehicles have multiple computers with multiple functions. The computers have the ability to communicate with each other. One computer receives some sensor inputs, and the signal is forwarded to other computers. If the signal is not received, this can be interpreted as an incorrect input and may cause output problems from the processor. Theft deterrent systems may cause stalling or no-start conditions that can be difficult to trace or lead a technician toward performing unnecessary testing and parts replacement. This may be especially true with aftermarket alarm systems. Often, when control modules that communicate with each other, like the PCM and theft deterrent module, are replaced, relearn procedures may need to be performed before normal operation will occur.

Other driveability concerns can occur from problems in the cruise control or traction control systems, such as surging or loss of power. Electrical interference may result from certain aftermarket-installed accessories, such as stereo amplifiers. Improper installation or wiring damage is also a concern when non-OEM components are installed.

All technicians will need to identify all systems that are used on a vehicle and what possible interaction they may have on the PCM. The vehicle service manual and service bulletins should be consulted when problems are suspected in interrelated systems.

▦ 11. Clear diagnostic trouble codes (DTCs), run all OBD-II monitors, and verify the repair.

Prior to the introduction of OBD-II, each manufacturer had its own method for erasing DTCs from the memory of a PCM. These procedures should always be followed. Normally, verification of the repair is done by operating the engine and the related system and checking to see if the operation triggered the DTC. If it did not, the problem was probably solved.

On OBD-II equipped vehicles, the fail records and the freeze frame data for the DTC that was diagnosed should be reviewed and recorded. Then use a scan tool's "clear DTCs" or "clear info" function to erase the DTCs from the PCM's memory. Operate the vehicle within the conditions noted in the fail records and/or the freeze frame data. Then, monitor the status information for the specific DTC until the diagnostic test associated with that DTC runs. A PCM runs certain monitors to test components such as the catalytic converter, the O2 sensor heater, and EGR systems. Enabling criteria must be met in order for the PCM to run a monitor. If all enabling criteria are not met, the monitor will not be run. Enabling criteria include temperature, vehicle speed, and elapsed time frames. Each monitor's enabling criteria is different.

Sample Preparation Exams

INTRODUCTION

Included in this section are a series of six individual preparation exams that you can use to help determine your overall readiness to successfully pass the Exhaust Systems (X1) ASE certification exam. Located in Section 7 of this book, you will find blank answer sheet forms you can use to designate your answers to each of the preparation exams. Using these blank forms will allow you to attempt each of the six individual exams multiple times without risk of viewing your prior responses.

Upon completion of each preparation exam, you can determine your exam score using the answer keys and explanations located in Section 6 of this book. Included in the explanation for each question is the specific task area being assessed by that individual question. This additional reference information may prove useful if you need to refer back to the task list located in Section 4 for additional support.

PREPARATION EXAM 1

1. Which of the following would be the normal maximum amount of warpage allowed for a cast iron cylinder head?

   A. 0.0001″
   B. 0.003″
   C. 0.010″
   D. 0.030″

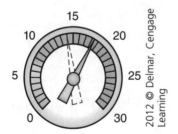

2. With the engine idling, a vacuum gauge connected to the intake manifold fluctuates as shown in the above illustration. These vacuum gauge fluctuations may be caused by:

   A. Late ignition timing
   B. Intake manifold vacuum leak at the throttle body
   C. A restricted exhaust system
   D. Sticky valve stems and guides

---

3. Which of the following is the most likely cause of low oil pressure at idle?

    A. Worn piston rings
    B. Worn piston pins
    C. Worn bearings
    D. Worn cam lobes

4. Which of the following tools would be used to bench check an ignition coil?

    A. Voltmeter
    B. Ohmmeter
    C. Ammeter
    D. Oscilloscope

5. A vehicle has low fuel pressure (5 psi) and will not start. Which of the following could be the cause?

    A. An open fuel pump relay
    B. An open fuel pump winding
    C. Low fuel level
    D. Low oil level

6. The last step on most diagnostic flow charts is:

    A. Verify the repair
    B. Perform the repair
    C. Identify the cause for concern
    D. Verify concern

7. The PCV valve is removed with the engine running. There is no change in engine operation. This would indicate:

    A. Normal engine operation
    B. A plugged valve
    C. A sticking valve
    D. A restricted hose

8. A surface comparator is used to:

    A. Compare surface warpage
    B. Compare surface roughness
    C. Compare head thickness
    D. Compare block thickness

9. All of the following is correct concerning diagnostic trouble codes (DTCs) EXCEPT:

    A. Some DTCs will not set until two complete drive cycles.
    B. Some DTCs are erased when the battery is disconnected.
    C. All DTCs are erased when the electronic control module (ECM) fuse is removed.
    D. Some DTCs will set on the first drive cycle.

10. Which of the following would most likely result from a restricted catalytic converter?

    A. Low power
    B. Increased engine idle RPM
    C. Increased engine vacuum
    D. Low combustion chamber temperatures

11. During a vehicle road test the vehicle blows blue smoke from the tailpipe. Which of the following could be the cause?

    A. Worn compression rings
    B. Worn oil control rings
    C. Worn intake valve seats
    D. Worn exhaust valve seats

12. A truck with a half a tank of gasoline has low power and stalls. The fuel tank is filled. The truck restarts, runs for an hour, and stalls again. After refueling, the sequence starts again. Which of the following could be the cause?

    A. The fuel pump is failing.
    B. The fuel pump relay is failing.
    C. The ECM is failing.
    D. The ignition control module is failing.

13. Cylinder head thickness is less than specified. The most likely cause is:

    A. Engine overheating
    B. Insufficient oil pressure
    C. Excessive milling
    D. Low engine operating temperature

14. The oil pressure gauge is reading lower than normal; what should the technician do first?

    A. Check the pressure with a master gauge
    B. Check the oil level
    C. Change the oil
    D. Change the oil filter

15. The acronym TSB means:

    A. Technical Sales Bulletin
    B. Trade School Books
    C. Technical Service Bulletin
    D. Trade Service Books

16. Which tool can be used to check catalytic converter efficiency?

    A. Compression gauge
    B. Cylinder leakage tester
    C. Infrared pyrometer
    D. Refractometer

17. All of the following can be used to measure a cylinder bore for wear EXCEPT:

    A. A dial bore gauge
    B. An inside micrometer
    C. An outside micrometer
    D. A telescoping gauge

18. An engine has set a misfire code for cylinder #5. The technician swaps the ignition coil on cylinders 5 and 6. Now there is a misfire code for cylinder #6. This most likely indicates a:

    A. Failed ECM
    B. Failed primary wiring harness
    C. Low compression on cylinder #5
    D. Failed ignition coil

19. Which of the following can be used to help indicate the source of an engine oil leak?

    A. Green light
    B. Black light
    C. Yellow light
    D. Strobe light

20. A fuel pump has a higher than normal current draw. Which of the following could be the cause?

    A. High resistance in the fuel pump relay
    B. Tight bushings in the fuel pump
    C. High resistance in the fuel pump wiring
    D. An open in the fuel pump relay

21. Which of the following would be used to measure valve guide inside diameter?

    A. Inside micrometer
    B. Depth micrometer
    C. Dial indicator
    D. Split ball gauge

22. When measuring the intake manifold vacuum at idle, the technician observes a regular downward pulsation, dropping from 17 in. Hg to 10 in. Hg, then immediately recovering to 17 in. Hg. Which of the following could be the cause?

    A. A leaking intake valve
    B. Worn piston rings
    C. Worn turbocharger bearings
    D. A leaking intake manifold

23. A DTC has been set for a shorted coolant temperature sensor. The scan tool is monitored while the sensor wiring harness is disconnected. The scan tool display does not change. Which of the following could be the cause?

    A. The sensor is shorted.
    B. The sensor is open.
    C. The wiring harness is open.
    D. The wiring harness is shorted.

24. Which of the following resistance readings would indicate an open spark plug wire?

    A.   0 ohms
    B.   100 ohms
    C.   5000 ohms
    D.   O.L.

25. Which of the following can be used to help indicate the source of an engine coolant leak?

    A.   Green light
    B.   Black light
    C.   Yellow light
    D.   Strobe light

26. A vehicle is equipped with a negative temperature coefficient coolant temperature sensor. Technician A says temperature sensor resistance will increase as temperature increases. Technician B says temperature sensor resistance will decrease as temperature decreases. Who is correct?

    A.   A only
    B.   B only
    C.   Both A and B
    D.   Neither A nor B

27. Which of the following can be used to help indicate the source of an engine vacuum leak?

    A.   Compression gauge
    B.   Cylinder leakage tester
    C.   Smoke machine
    D.   Oscilloscope

28. An engine has a high idle concern. When the PCV hose is crimped shut, the engine RPM drops to normal. Which of the following is indicated?

    A.   A restricted positive crankcase ventilation (PCV) system
    B.   A stuck open idle air control valve
    C.   A stuck closed idle air control valve
    D.   A failed open PCV valve

29. Which of the following can be used to help indicate the source of an evaporative emissions (EVAP) system leak?

    A.   Compression gauge
    B.   Cylinder leakage tester
    C.   Smoke machine
    D.   Oscilloscope

30. Which of the following is the most common method for installing in-block camshaft bearings?

    A.   Driven in
    B.   Heated, then dropped in
    C.   Cooled, then dropped in
    D.   Cast in

31. A truck has a single cylinder misfire. There is no spark at the spark plug. There is spark at the plug wire. Which of the following could be the cause of the misfire?

    A. Low fuel pressure

    B. Low compression

    C. Worn cam lobe

    D. Failed spark plug

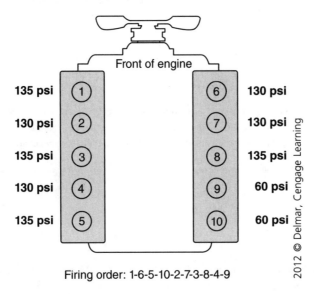

Firing order: 1-6-5-10-2-7-3-8-4-9

32. Compression test results for an engine are shown in the above illustration. Technician A says this could be caused by a loose timing chain. Technician B says a leaking head gasket could cause this. Who is correct?

    A. A only

    B. B only

    C. Both A and B

    D. Neither A nor B

33. Short term fuel trim indicates the ECM is adding extra fuel. This could be caused by:

    A. A stuck open fuel injector

    B. A leaking exhaust manifold

    C. A stuck closed fuel pressure regulator

    D. A leaking exhaust tailpipe

34. Valve springs should be measured for all of the following EXCEPT:

    A. Squareness

    B. Free length

    C. Tension

    D. Radius

35. While monitoring the inlet air temperature sensor value with a scan tool, the technician finds that the value fluctuates when the throttle is opened and closed rapidly. This indicates:

   A. The temperature sensor is operating normally.
   B. The temperature sensor is shorted.
   C. The temperature sensor is open.
   D. The air filter is restricted.

36. An engine has a light ticking sound. Engine oil pressure is normal. Which of the following could be the cause?

   A. Worn rod bearings
   B. Worn main bearings
   C. Valves adjusted too loose
   D. Valves adjusted too tight

37. There is a light that resembles a gas cap illuminated on the dash. The technician should:

   A. Check the fuel level in the fuel tank
   B. Crain the fuel tank
   C. Inspect the gas cap
   D. Check fuel pump pressure

38. While monitoring the mass airflow sensor value with a scan tool, the technician finds that the value fluctuates when the throttle is opened and closed rapidly. This indicates:

   A. The mass airflow sensor is operating normally.
   B. The mass airflow sensor is shorted.
   C. The mass airflow sensor is open.
   D. The sensor is dirty.

39. A vehicle has poor fuel economy. Technician A says the cause could be a restricted air filter. Technician B says the cause could be a seized fan clutch. Who is correct?

   A. A only
   B. B only
   C. Both A and B
   D. Neither A nor B

40. Engine balance shafts can be driven by all of the following EXCEPT:

   A. Gears
   B. Belts
   C. Chains
   D. Electric motors

41. An engine has a rough unstable idle and poor acceleration. Technician A says a stuck open exhaust gas recirculation (EGR) valve could be the cause. Technician B says a stuck closed PCV valve could be the cause. Who is correct?

   A. A only
   B. B only
   C. Both A and B
   D. Neither A nor B

42. The specification for the cylinder bore on a given cast iron engine is 4.0125″. Listed below are the measurements for the engine:

Cylinder #1	4.0126″	Cylinder #5	4.0126″
Cylinder #2	4.0125″	Cylinder #6	4.0126″
Cylinder #3	4.0126″	Cylinder #7	4.0125″
Cylinder #4	4.0127″	Cylinder #8	4.0127″

    Which of the following would be the most likely repair procedure?

    A. Bore the block
    B. Sleeve the block
    C. Reuse the block as is
    D. Scrap the block

43. A truck will not start. When the fuel pump relay is jumped, the truck starts. Which of the following could be the cause?

    A. A failed fuel pump
    B. A failed ignition control module
    C. A failed fuel injector control module
    D. A failed fuel pump relay

44. An engine has a single cylinder misfire; there is a popping sound from the exhaust when the misfire occurs. Which of the following could be the cause?

    A. A hole in a piston
    B. A leaking intake gasket
    C. A leaking thermostat housing gasket
    D. A burnt exhaust valve

45. An engine has suddenly lost oil pressure. Which of the following could be the cause?

    A. The main bearings are worn.
    B. The rod bearings are worn.
    C. The oil pump pickup tube has come loose.
    D. The piston rings have broken.

46. All of the following can be used to test an accelerator pedal position (APP) sensor EXCEPT:

    A. Oscilloscope
    B. Ohmmeter
    C. Ammeter
    D. Scan tool

47. While diagnosing a DTC, the technician finds a technical service bulletin that directly conflicts with service information for the vehicle. Which of the following should the technician do?

    A. Follow the original service literature.
    B. Follow the technical service bulletin.
    C. Tell the customer to take the vehicle elsewhere.
    D. Replace the part and see if that fixes the vehicle.

48. An engine has low power. One cylinder has 40 percent leakage during a cylinder leakage test. All other cylinders have 15 percent leakage. Which of the following is true?

    A. This is a normal reading.
    B. This can be caused be retarded ignition timing.
    C. This can be caused by retarded camshaft timing.
    D. This can be caused by broken compression rings.

49. A vehicle has low power. Technician A says the cause could be retarded ignition timing. Technician B says the cause could be low ignition coil secondary voltage.
    Who is correct?

    A. A only
    B. B only
    C. Both A and B
    D. Neither A nor B

50. A truck dies while it is being driven; it will crank but not start. There are no DTCs. Which of the following should the technician check first?

    A. Throttle position sensor (TPS)
    B. APP
    C. EGR
    D. Fuel Level

PREPARATION EXAM 2

1. The specification for the cylinder bore on a given cast iron engine is 4.0225″. Listed below are the measurements for the engine:

Cylinder #1 4.0285″	Cylinder #5 4.0246″
Cylinder #2 4.0265″	Cylinder #6 4.0266″
Cylinder #3 4.0276″	Cylinder #7 4.0275″
Cylinder #4 4.0267″	Cylinder #8 4.0307″

    Which of the following would be the most likely repair procedure?

    A. Bore the block
    B. Sleeve the block
    C. Reuse the block as is
    D. Scrap the block

2. The first diagnostic step on most flow charts is:

    A. Verify the repair.
    B. Perform the repair.
    C. Identify the cause for concern.
    D. Verify concern.

3. Which of the following engine lubricating oils would typically be used in a gasoline-powered truck engine?

    A. 0W-20

    B. SAE40

    C. 10W-30

    D. 15W-40

4. The normal resistance for the primary windings of an ignition coil is:

    A. 1 ohm

    B. 100 ohms

    C. 1,000 ohms

    D. 10,000 ohms

5. An engine has a single cylinder misfire and there is a regularly occurring popping sound in the intake manifold. Which of the following could be the cause?

    A. A broken exhaust pushrod

    B. A broken intake pushrod

    C. A faulty intake manifold pressure sensor

    D. A faulty coolant temperature sensor

6. There is blue smoke coming from the exhaust of a gasoline-powered truck. Which of the following could be the cause?

    A. A leaking radiator

    B. A leaking oil cooler

    C. Worn piston rings

    D. Worn cam lobes

7. A truck has low power. The fuel pressure is checked and is found to be lower than specification. Which of the following could be the cause?

    A. A stuck closed pressure regulator

    B. A stuck closed fuel injector

    C. Low manifold vacuum

    D. A restricted fuel filter

8. A vehicle runs acceptably while traveling at road speed but stalls when coming to a stop, is difficult to restart, and will not idle. Which of the following could be the cause?

    A. Low fuel pressure

    B. A restricted fuel filter

    C. A stuck open exhaust gas recirculation (EGR) valve

    D. A stuck closed EGR valve

9. An engine is equipped with an AC voltage generator style crankshaft position sensor (CKP). This sensor can be bench checked with a:

    A. Dwell meter

    B. Ohmmeter

    C. Oscilloscope

    D. Ammeter

10. There is black smoke coming from the exhaust of a gasoline-powered truck. Which of the following could be the cause?

   A. A leaking fuel injector
   B. A leaking head gasket
   C. A leaking exhaust valve
   D. A leaking exhaust manifold

11. An engine has a surging problem at highway speeds. Engine operation is normal at idle and low speeds. Technician A says there may be a high resistance across the fuel pump relay. Technician B says that the inertia switch may have high resistance. Who is correct?

   A. A only
   B. B only
   C. Both A and B
   D. Neither A nor B

12. There is the noticeable smell of exhaust in the cab and a backfire on engine deceleration. Which of the following could be the cause?

   A. A cracked intake manifold
   B. A leaking coolant temperature sensor
   C. A cracked exhaust manifold
   D. An internally leaking EGR valve

13. A fuel pump has been replaced because of high current draw and low fuel pressure. In addition to replacing the fuel pump, the technician should also inspect:

   A. Voltage drop across the fuel pump electrical relay
   B. Applied voltage to the electronic control module (ECM)
   C. Voltage drop across the fuel injector electrical relay
   D. Applied voltage to the ignition control module

14. Which tool would most likely be used to measure camshaft end-play on an in-block camshaft?

   A. Feeler gauge
   B. Dial indicator
   C. Inside micrometer
   D. Depth micrometer

15. Which of the following would be the most normal intake manifold vacuum reading at idle?

   A. 5 in. Hg
   B. 10 in. Hg
   C. 17 in. Hg
   D. 27 in. Hg

16. Technician A says spark plugs should be installed using a torque wrench. Technician B says spark plugs must be removed using a torque wrench. Who is correct?

   A. A only
   B. B only
   C. Both A and B
   D. Neither A nor B

17. Which of the following would be used to measure cylinder head warpage?

    A. Micrometer
    B. Dial indicator
    C. Caliper and feeler gauge
    D. Straightedge and feeler gauge

18. A truck backfires on deceleration. Technician A says a faulty secondary air injection valve could be the problem. Technician B says an exhaust leak could be the problem. Who is correct?

    A. A only
    B. B only
    C. Both A and B
    D. Neither A nor B

19. A truck is being diagnosed for a no-start concern. Technician A says that the level of fuel in the tank should be checked. Technician B says injector pulse width (PW) should be checked. Who is right?

    A. A only
    B. B only
    C. Both A and B
    D. Neither A nor B

20. Which of the following can be used to help indicate the source of an engine oil leak?

    A. Smoke machine
    B. Pressure gauge
    C. Vacuum gauge
    D. Oscilloscope

21. An engine has high hydrocarbon (HC) and carbon monoxide (CO) emissions. All of these defects could cause this problem EXCEPT:

    A. A leaking fuel injector
    B. An engine coolant temperature (ECT) sensor with low resistance
    C. An ECT sensor with high resistance
    D. A defective manifold absolute pressure (MAP) sensor

22. Technician A says a plugged catalytic converter can be caused by a stuck open fuel injector. Technician B says a plugged catalytic converter can be caused by a misfiring cylinder. Who is correct?

    A. A only
    B. B only
    C. Both A and B
    D. Neither A nor B

23. The crankshaft main bearing journal specification is 2.569″. The actual measurements are listed below:

    Measurement #1 2.568″

    Measurement #2 2.567″

    Measurement #3 2.567″

    Measurement #4 2.566″

    Which of the following would be the most likely service procedure?

    A. Reuse the crankshaft as is.
    B. Turn the crankshaft 0.010″.
    C. Turn the crankshaft 0.005″.
    D. Weld the crankshaft and turn to original dimensions.

24. Which of these would be LEAST LIKELY to cause a no-start condition?

    A. A dirty air filter
    B. A restricted fuel filter
    C. A faulty oxygen sensor
    D. A faulty CKP

25. While using a vacuum gauge, the technician finds the intake manifold vacuum is low and steady on a gasoline engine. Which of the following is the most likely cause?

    A. A leaking exhaust valve
    B. A leaking intake valve
    C. A restricted exhaust
    D. A restricted air filter

26. A vehicle is equipped with an ignition control module and a fuel injector control module. The ignition control module has shorted. Which of the following is the most likely cause of the failure?

    A. A shorted fuel injector
    B. An open fuel injector
    C. An open coil pack
    D. A shorted coil pack

27. The evaporative emissions control (EVAP) system is inoperative. Looking at the system monitors on the scan tool, the technician finds that the EVAP system monitor never runs. Technician A says a faulty engine temperature sensor signal could cause this. Technician B says a leaking vacuum hose at the EVAP solenoid could cause this. Who is correct?

    A. A only
    B. B only
    C. Both A and B
    D. Neither A nor B

28. An intake manifold vacuum test is being performed on a gasoline-powered V8 truck engine. Which of the following would most likely be identified by a vacuum test?

    A. A fouled spark plug on cylinder #6
    B. A leaking valve on cylinder #5
    C. A restricted exhaust
    D. A loose piston pin

29. All of the following are true regarding the use of a scan tool EXCEPT:

    A. The engine should be warmed to determine if the PCM enters closed loop.
    B. The scan tool power adapter should be connected to a 12 V power supply.
    C. The PCM fuse should be removed prior to connecting to the scan tool.
    D. The year, model, and engine may be read automatically from the PCM.

30. A technician is using a vacuum gauge to measure cranking vacuum. Which of the following would be the most normal reading?

    A. 1 in. Hg
    B. 4 in. Hg
    C. 10 in. Hg
    D. 18 in. Hg

31. A truck requires extended cranking before it will start. Which of the following could be the cause?

    A. A faulty fuel pump relay
    B. Low battery voltage
    C. Carbon buildup on the pistons
    D. A restricted exhaust

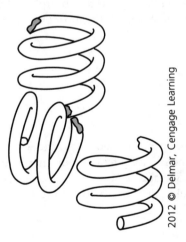

2012 © Delmar, Cengage Learning

32. Which of the following can cause the valve spring damage pictured above?

    A. Excessive idling
    B. A stuck open thermostat
    C. Engine over-speeding
    D. Incorrect piston-to-cylinder wall clearance

33. Technician A says a vacuum gauge can be used to help determine if the catalytic converter is clogged. Technician B says a fuel pressure gauge can be used to help determine if the catalytic converter is clogged. Who is correct?

A. A only
B. B only
C. Both A and B
D. Neither A nor B

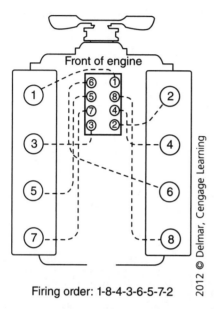

Firing order: 1-8-4-3-6-5-7-2

34. A vehicle has a firing order of 1-8-4-3-6-5-7-2, as seen in the cylinder arrangement pictured above. A diagnostic trouble code (DTC) has been set for a misfire on cylinders #5 and #7. All other cylinders are performing normally. Which of the following could be the cause?

A. Low fuel pump pressure
B. A failed head gasket
C. A worn exhaust cam lobe
D. A worn intake cam lobe

35. The intake manifold vacuum is being measured during a road test. The vacuum gauge continues to drop the longer the vehicle is driven, and eventually the engine fails to accelerate and stalls. Which of the following could be the cause?

A. A restricted EGR passage
B. A restricted exhaust manifold
C. A leaking exhaust manifold
D. A leaking intake manifold

36. The engine is overheating. Which of the following is the most likely cause?

A. Engine thermostat stuck open
B. Engine cooling fan stuck on
C. Engine thermostat stuck closed
D. Overtightened water pump belt

37. Technician A says that a defective EGR valve may cause an engine to be hard to start. Technician B says that a defective ECT sensor may cause higher than normal emissions. Who is correct?

    A. A only
    B. B only
    C. Both A and B
    D. Neither A nor B

38. An engine has spun a rod bearing. Besides repairing the crankshaft, what other service procedure must be performed?

    A. Bore the cylinders
    B. Machine the connecting rod
    C. Mill the head
    D. Deck the block

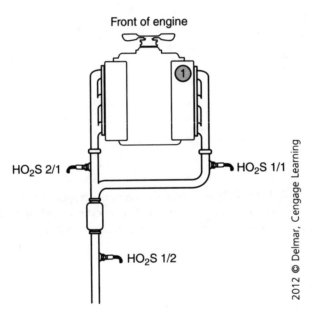

39. Refer to the above illustration. Which of the following can occur if the oxygen sensors HO$_2$S 1/1 and HO$_2$S 2/1 fail?

    A. The GPS system will not function correctly.
    B. The engine will run colder than normal.
    C. The system will never go to closed loop.
    D. The engine will not start.

40. A cylinder power balance test is being performed on a gasoline-powered V8 truck engine. One cylinder reads lower than the others. Which of the following could be the cause?

    A. A plugged fuel injector
    B. A plugged exhaust system
    C. A faulty oxygen sensor
    D. A faulty oil pressure sensor

41. A customer has had two failures of the EVAP system charcoal canister. Both canisters were flooded. All of the following could be the cause EXCEPT:

    A.   Customer fueling habits
    B.   A failed canister purge valve
    C.   A plugged EGR passage
    D.   A plugged vacuum line

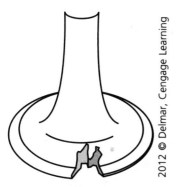

42. All the exhaust valves in an engine are damaged, as pictured in the above illustration. Which of the following could be the cause?

    A.   Adjusting the valves too tight
    B.   Adjusting the valves too lose
    C.   A faulty throttle position sensor (TPS)
    D.   A leaking fuel injector

43. Digital multi-meters (DMM) can be used on all the following scales EXCEPT:

    A.   Milli-watts
    B.   Amps
    C.   Volts
    D.   Ohms

44. During a cylinder leakage test, shop air pressure should be set to:

    A.   25 psi
    B.   50 psi
    C.   100 psi
    D.   150 psi

45. Which of the following would most likely set a multiple cylinder misfire DTC?

    A.   Low oil level voltage
    B.   Low fuel pressure
    C.   High fuel pressure
    D.   A faulty post-catalyst oxygen sensor

46. Technician A says that an APP sensor can be checked with the DMM on the resistance scale. Technician B says that an APP sensor can be checked with the meter set on the AC voltage scale. Who is correct?

    A. A only
    B. B only
    C. Both A and B
    D. Neither A nor B

47. Engine coolant test strips can measure all of the following EXCEPT:

    A. Coolant temperature
    B. Coolant freeze protection
    C. Coolant pH
    D. Coolant acidity

48. Air is heard escaping the tailpipe during a cylinder leakage test. Which of the following could be the cause?

    A. A leaking intake valve
    B. A leaking exhaust valve
    C. A leaking head gasket
    D. A leaking exhaust manifold

49. A coil-on-plug (COP) ignition coil has low secondary voltage output. All of the following could be the cause EXCEPT:

    A. Low primary voltage
    B. High resistance in the primary windings
    C. A failed ignition coil
    D. A failed crankshaft sensor

50. What are valve spring shims used for?

    A. Correct weak valve springs
    B. Correct valve spring installed height
    C. Correct valve spring free length
    D. Correct valve spring squareness

PREPARATION EXAM 3

1. Technician A says that an exhaust gas recirculation (EGR) position sensor may be tested with either a voltmeter or an oscilloscope. Technician B says that if the powertrain control module (PCM) does not set an EGR position sensor fault code, there is no problem with the sensor. Who is correct?

    A. A only
    B. B only
    C. Both A and B
    D. Neither A nor B

2. A customer has a low oil pressure concern. After verifying the complaint, what should the technician do next?

   A. Replace the gauge in the dash.
   B. Replace the oil pressure sending unit.
   C. Measure the engine oil pressure with a master gauge.
   D. Check for proper engine oil level.

3. The normal resistance when measuring the primary to secondary windings of an ignition coil would be:

   A. 1 ohm
   B. 100 ohms
   C. 1,000 ohms
   D. 10,000 ohms

4. A vehicle's electronic control module (ECM) must be replaced. Technician A says it is good practice to use a ground strap to ensure that there is no static discharge to the electronic circuit. Technician B says disconnecting the vehicle battery and touching bare metal on the vehicle prior to replacement will protect the vehicle from static discharge.
   Who is right?

   A. A only
   B. B only
   C. Both A and B
   D. Neither A nor B

5. One intake manifold runner has a vacuum leak. Which of the following would most likely occur?

   A. The engine would not start.
   B. The engine would have excessive power.
   C. The engine would not shut off.
   D. The engine idle RPM would increase.

6. Plastigauge® is typically used to:

   A. Measure piston-to-cylinder wall clearance
   B. Measure bearing clearance
   C. Measure camshaft end-play
   D. Measure crankshaft end-play

7. The valves need to be adjusted on an overhead camshaft (OHC) engine that uses a shim style adjustment method. Technician A says the shim should be removed, then the clearance measured. Technician B says if the clearance is out of specification, the shim should be removed and ground to the appropriate dimension. Who is correct?

   A. A only
   B. B only
   C. Both A and B
   D. Neither A nor B

8. A truck has a diagnostic trouble code (DTC) for a leaking evaporative emissions (EVAP) system. Which of the following tools would be used to locate the leak?

   A. A nitrogen machine
   B. A smoke machine
   C. A cooling system pressure tester
   D. A compression gauge

9. A refractometer is used to measure coolant:

   A. Temperature
   B. Freeze protection
   C. pH
   D. Acidity

10. A truck is brought to the shop with a fuel-related DTC. Technician A says all codes should be recorded; then, they should be erased to see if any are active codes. Technician B says some scan tools list active and inactive for ease of interpretation. Who is correct?

    A. A only
    B. B only
    C. Both A and B
    D. Neither A nor B

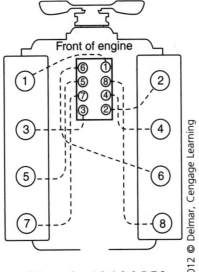

Firing order: 1-8-4-3-6-5-7-2

2012 © Delmar, Cengage Learning

11. Refer to the above illustration. A vehicle has a firing order of 1-8-4-3-6-5-7-2. A DTC has been set for a misfire on cylinders #8 and #5. All other cylinders are performing normally. Which of the following could be the cause?

    A. Low fuel pressure
    B. A failed head gasket
    C. A failed ignition coil
    D. A worn intake cam lobe

12. Cylinder head bolt torquing sequence usually begins:

    A. In the middle
    B. At the exhaust side
    C. At the intake side
    D. At the end

13. A truck has a DTC for a non-functioning catalytic converter. While comparing the oxygen sensors 2/1 and 2/2, the technician finds they are extremely similar in their voltage and response. Technician A says this indicates a non-functioning catalytic converter. Technician B says this indicates a non-functioning oxygen sensor 2/2. Who is correct?

    A. A only
    B. B only
    C. Both A and B
    D. Neither A nor B

14. A technician is preparing to install a backpressure gauge to check the catalytic converter for restriction. Which of the following would be the correct location?

    A. In spark plug hole #6
    B. In spark plug hole #8
    C. In oxygen sensor hole 1/1
    D. In oxygen sensor hole 2/1

15. All of the following could cause a single cylinder misfire EXCEPT:

    A. A fouled spark plug
    B. A dirty throttle plate
    C. A stuck open fuel injector
    D. A stuck closed fuel injector

16. During a cylinder leakage test, bubbles are found in the radiator. Which of the following is indicated?

    A. A restricted radiator
    B. A restricted head gasket
    C. A leaking radiator
    D. A leaking head gasket

17. The color of Plastigauge determines its:

    A. Diameter
    B. Length
    C. Temperature
    D. Age

18. A DTC has been set for the EGR system. Technician A says plugged passage tubes connecting the EGR pressure sensor could be the cause. Technician B says a failed EGR pressure sensor could be the cause. Who is correct?

    A. A only
    B. B only
    C. Both A and B
    D. Neither A nor B

19. Which of the following is a normal compression reading during a compression test?

    A. 50 psi
    B. 75 psi
    C. 125 psi
    D. 225 psi

20. A coil on plug ignition coil has failed and must be replaced. Technician A says the ignition coil on the companion cylinder must also be replaced. Technician B says the fuel injector must also be replaced. Who is correct?

    A. A only
    B. B only
    C. Both A and B
    D. Neither A nor B

21. Which of the following would be performed during a power balance test?

    A. Remove the spark from a cylinder.
    B. Remove the coolant from a cylinder.
    C. Remove the intake air from a cylinder.
    D. Remove the exhaust from a cylinder.

22. All cylinders read lower than normal on a cranking compression test. Which of the following could be the cause?

    A. Excessive cranking speed
    B. A leaking intake valve
    C. A leaking exhaust valve
    D. An out-of-time camshaft

23. An exhaust manifold is leaking at the cylinder head. Technician A says the manifold must be replaced. Technician B says the cylinder head must be replaced. Who is correct?

    A. A only
    B. B only
    C. Both A and B
    D. Neither A nor B

24. Piston ring end gap is measured:

    A. With the rings on the piston
    B. With the rings in the cylinder bore
    C. With a dial indicator
    D. After the engine is assembled

25. A customer is concerned that coolant must be added to the cooling system. The technician finds no evidence of an external leak. A cylinder leakage test reveals bubbles in the coolant. Which condition is indicated?

    A. Leaking radiator
    B. Leaking oil cooler
    C. Leaking head gasket
    D. Leaking throttle plate gasket

26. Where would a technician locate the most recent software upgrades for an engine ECM?

    A. Service manuals

    B. CD-ROMs

    C. The original equipment manufacturer (OEM) service website

    D. Trade publications

27. The air intake system is being inspected for leaks. Technician A says a smoke machine may be used. Technician B says starting fluid may be used. Who is correct?

    A. A only

    B. B only

    C. Both A and B

    D. Neither A nor B

28. A vehicle has the check engine light on. The cruise control, automatic door locks, and speedometer are not functioning. Which of the following is the most likely cause?

    A. A faulty PCM

    B. A faulty oxygen sensor

    C. A faulty vehicle speed sensor

    D. A faulty wheel speed sensor

29. An engine with an OHC has had the deck milled. Technician A says this can affect camshaft timing. Technician B says this can affect fuel pressure. Who is correct?

    A. A only

    B. B only

    C. Both A and B

    D. Neither A nor B

30. An engine has a knock that is not load dependent and can only be heard at 1000 rpm, 1700 rpm, 2400 rpm, and 3100 rpm. Which of the following is the most likely cause?

    A. Rod bearing

    B. Main bearing

    C. Piston pin

    D. Vibration damper

31. The acronym CMP is used for:

    A. Camshaft Position Sensor

    B. Crankcase Pressure Sensor

    C. Crankshaft Position Sensor

    D. Cylinder Pressure Sensor

32. An engine has an oil leak. Technician A says fluorescent dye can be used to locate the source of the leak. Technician B says smoke can be used to locate the source of the leak. Who is correct?

    A. A only

    B. B only

    C. Both A and B

    D. Neither A nor B

33. Which of the following would be a common metric bolt designation?

    A. M8 × 1.0 mm
    B. M8 × 18 tpi
    C. 3/8 × 18 tpi
    D. 3/8 × 1.5 mm

34. An engine which has high oil consumption also has high crankcase pressure. Which of the following is the most likely cause?

    A. Worn intake valve guides
    B. Worn piston rings
    C. Worn exhaust valve guides
    D. Worn camshaft lobes

35. A vacuum-controlled fuel pressure regulator is being tested. What is the correct procedure?

    A. With the engine shut off, monitor fuel pressure while removing the vacuum hose.
    B. Remove the regulator and submerge it in fuel.
    C. Remove the regulator from the engine and apply vacuum.
    D. With the engine running, monitor fuel pressure while removing the vacuum hose.

36. There is a DTC set for an inoperative EGR valve. When the technician opens the EGR valve manually with the engine running, the engine stumbles and dies. Which of the following could be the cause of the inoperative EGR valve code?

    A. Restricted EGR passages
    B. Failed EGR pressure sensor
    C. Clogged EVAP canister converter
    D. Punctured catalytic converter

37. Technician A says a defective ECT sensor may cause the engine to be hard to start. Technician B says a defective ECT sensor may cause the engine to fail an emissions test. Who is correct?

    A. A only
    B. B only
    C. Both A and B
    D. Neither A nor B

38. An engine has normal intake manifold vacuum at idle. When the vehicle is driven, the intake manifold vacuum continually gets lower, and eventually the engine stalls. Which test should the technician perform next?

    A. Exhaust restriction
    B. Cranking compression
    C. Running compression
    D. Air filter restriction

39. Technician A says fuel injectors can be cleaned on the engine. Technician B says fuel injectors can be cleaned on a test bench. Who is correct?

    A. A only
    B. B only
    C. Both A and B
    D. Neither A nor B

40. During a power balance test, two adjacent cylinders are found to have low power. Which of the following could be the cause?

    A. A restricted fuel filter
    B. A restricted air filter
    C. Switched spark plug wires
    D. Switched oxygen sensor connectors

41. A power balance test is being performed on a gasoline-powered V10 truck engine. The front six cylinders have good power output. The rear four cylinders have less than adequate power output. Which of the following could be the cause?

    A. A restricted air filter
    B. Low engine oil level
    C. Low transmission fluid level
    D. Restricted coolant passages

42. Technician A says that closed-loop operation occurs when the engine is at operating temperature, and the PCM uses the information from the oxygen sensor to manage the air/fuel ratio. Technician B says that while in closed-loop operation, injector pulse width (PW) is measured in seconds. Who is correct?

    A. A only
    B. B only
    C. Both A and B
    D. Neither A nor B

43. An engine has low power. A cranking compression test reveals two adjacent cylinders have low compression. Which additional test will best help isolate the problem?

    A. Intake manifold vacuum
    B. Exhaust restriction
    C. Cylinder leakage
    D. Running compression

44. A cylinder leakage test is being performed on a truck. Which of the following would indicate an acceptable amount of cylinder leakage?

    A. 15 percent
    B. 30 percent
    C. 45 percent
    D. 60 percent

45. Technician A says a surface comparator is used to determine the surface hardness of metal. Technician B says a surface comparator can be used to check the machined surface of a cylinder block. Who is correct?

    A. A only
    B. B only
    C. Both A and B
    D. Neither A nor B

46. There is no spark at any of the spark plugs on a truck equipped with a coil-on-plug ignition system. Which of the following could be the cause?

    A. A faulty mass airflow sensor
    B. A faulty ignition coil
    C. A failed CKP
    D. A failed EGR pressure sensor

47. A truck has a factory fill of orange DEX-COOL® as the coolant. Technician A says the coolant can be topped off with green ethylene glycol. Technician B says the coolant can be topped off with European Pink. Who is correct?

    A. A only
    B. B only
    C. Both A and B
    D. Neither A nor B

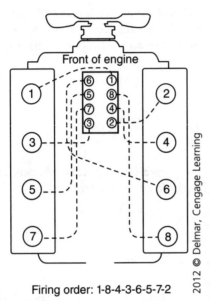

Firing order: 1-8-4-3-6-5-7-2

2012 © Delmar, Cengage Learning

48. Refer to the above illustration. A cylinder power balance test has been performed on an engine. The firing order of the engine is 1-8-4-3-6-5-7-2. Cylinders #1 and #8 show lower RPM drop than the others. Which of the following is the most likely cause?

    A. Restricted cooling system
    B. Incorrect spark plug wire routing
    C. Leaking intake valves
    D. Leaking compression rings

49. Spark plugs should be installed with a:

    A. 3/8″ dr. air impact wrench
    B. 3/8″ dr. butterfly
    C. 3/8″ dr. electric impact
    D. Torque wrench

50. OBD-II provides for standardization of all of the following EXCEPT:

    A. Communication protocols
    B. Diagnostic connectors
    C. Fuel calibrations
    D. DTCs

PREPARATION EXAM 4

1. Technician A says an enhanced evaporative emissions (EVAP) system must be able to detect a leak resulting from a 0.020 inch gap in the system. Technician B says the enhanced system must be able to detect a loose gas cap. Who is correct?

    A. A only
    B. B only
    C. Both A and B
    D. Neither A nor B

2. An engine has a low power complaint. A power balance test has been performed, and all the cylinders have about the same amount power output. Which of the following could be the cause of the low power concern?

    A. A restricted exhaust system
    B. A shorted coil pack
    C. A shorted spark plug wire
    D. A burnt exhaust valve

3. All of the following can be used to test a throttle position sensor (TPS) EXCEPT:

    A. Oscilloscope
    B. Ohmmeter
    C. Ammeter
    D. Scan tool

4. The acronym CKP is used for:

    A. Camshaft Position Sensor
    B. Crankcase Pressure Sensor
    C. Cylinder Pressure Sensor
    D. Crankshaft Position Sensor

5. Technician A says that prior to adjusting the throttle plates on a multi-port injected engine throttle body, the throttle plates and bore should be cleaned if dirty or varnished. Technician B says the TPS needs to be readjusted on some vehicles after the throttle plate angle adjustment is complete. Who is correct?

    A. A only
    B. B only
    C. Both A and B
    D. Neither A nor B

6. To check coil available voltage output, the technician should:

   A. Disconnect the fuel pump power lead.

   B. Disconnect the plug wire at the plug and ground it.

   C. Disconnect the coil wire and ground it.

   D. Conduct the test using a suitable spark tester that requires 25 kV.

7. The radiator is low on coolant, and the recovery bottle is overfull. Which of the following could be the cause?

   A. Tight water pump bearings

   B. Leaking radiator cap seal

   C. Incorrect cooling fan installed

   D. Incorrect fan clutch installed

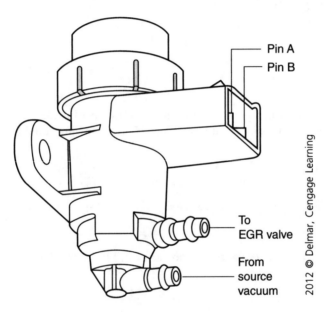

Pin A
Pin B
To EGR valve
From source vacuum

2012 © Delmar, Cengage Learning

8. A diagnostic trouble code (DTC) for an exhaust gas recirculation (EGR) problem is retrieved. A reading of infinite between Pin A and Pin B of the EGR vacuum regulator valve shown in the above illustration could mean:

   A. Nothing.

   B. The regulator is defective.

   C. The EGR valve is defective.

   D. The manifold absolute pressure (MAP) sensor is defective.

9. Technician A says a defective ECT sensor may cause a no-start condition. Technician B says a defective ECT sensor may cause the customer to have a transmission shift concern. Who is correct?

   A. A only

   B. B only

   C. Both A and B

   D. Neither A nor B

10. While adjusting valves with mechanical lifters, Technician A says when the valve clearance is checked on a cylinder, the piston should be positioned at top dead center (TDC) on the compression stroke. Technician B says the valves should be open when adjusted. Who is correct?

    A. A only
    B. B only
    C. Both A and B
    D. Neither A nor B

11. While performing a cylinder leakage test, air is heard escaping from the adjacent cylinder. This could be caused by:

    A. Failure to position the engine correctly
    B. A leaking exhaust valve
    C. A leaking head gasket
    D. A leaking intake valve

12. While performing a scan test on an OBD-II certified vehicle, a DTC P1336 is retrieved. Technician A says that a first digit, P, means the code is a generic trouble code. Technician B says that a second digit, 1, means the code is a manufacturer specific code. Who is correct?

    A. A only
    B. B only
    C. Both A and B
    D. Neither A nor B

13. An engine produces a bottom-end knock when started. Which of the following could be the cause?

    A. A bent pushrod
    B. One or more collapsed lifters
    C. Lack of oil pressure to the valve train
    D. Worn main bearings

14. A truck with port fuel injection is running roughly. A lab scope shows each injector waveform to be identical, except for one that has a considerably shorter voltage spike than the others. Which of the following is the most likely cause?

    A. Faulty powertrain control module (PCM)
    B. Open connection at the injector
    C. Shorted injector winding
    D. Low charging system voltage

15. An air pump is being replaced. Technician A says new air pumps will usually come with a new switching valve. Technician B says that the new air pump may not come with a new pulley. Who is correct?

    A. A only
    B. B only
    C. Both A and B
    D. Neither A nor B

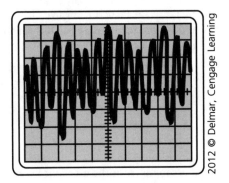

2012 © Delmar, Cengage Learning

16. Based on the oxygen sensor waveform shown in the above illustration, all the following are true EXCEPT:

    A. A lean biased condition is represented.
    B. A DTC may be recorded in the PCM.
    C. A rich biased condition is represented.
    D. The oxygen sensor is functioning.

17. Technician A says a power balance test can be performed with a scan tool on some trucks. Technician B says a power balance test can be performed using a test light on some trucks. Who is correct?

    A. A only
    B. B only
    C. Both A and B
    D. Neither A nor B

18. While replacing an air-injection manifold, which of the following should be done?

    A. Remove the exhaust manifold from the vehicle, then remove the air-injection manifold.
    B. Remove the air pump from the vehicle, and then remove the air-injection manifold.
    C. Use high-temperature silicone in place of the gasket.
    D. Apply penetrating oil to the nuts before removal.

19. Technician A says a high resistance connection at the alternator electrical connector could cause excessive heat buildup in the connector. Technician B says a connector that is too tight can cause heat distortion. Who is correct?

    A. A only
    B. B only
    C. Both A and B
    D. Neither A nor B

20. The LEAST LIKELY condition caused by a defective MAP sensor would be:

    A. A rich or lean air/fuel ratio
    B. Engine surging
    C. Excess fuel consumption
    D. Excessive idle speeds

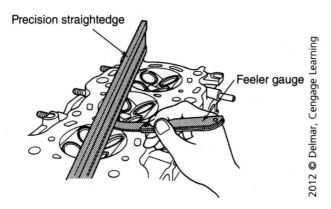

Precision straightedge

Feeler gauge

2012 © Delmar, Cengage Learning

21. In the above illustration, when measuring the cylinder head for warpage with a straightedge, the feeler gauge measurement is 0.012 in. (0.3 mm). Technician A says that measurement is higher than specified. Technician B says the measurement indicates that a thicker head gasket must be installed. Who is correct?

    A. A only
    B. B only
    C. Both A and B
    D. Neither A nor B

22. A charging system fuse link is being replaced. Technician A says the new fuse link should be soldered in place. Technician B says the new fuse link should be one size larger than the original to prevent it from burning again. Who is correct?

    A. A only
    B. B only
    C. Both A and B
    D. Neither A nor B

23. All of the following are true when testing a Magnetic Pulse Generator pickup coil EXCEPT:

    A. The pickup coil should be within the manufacturer's specified resistance value.
    B. An erratic reading while wiggling the pickup coil wires indicates that the pickup coil is intermittent.
    C. A resistance reading above manufacturer's specification indicates an open pickup coil.
    D. A resistance reading below manufacturer's specification indicates an open pickup coil.

24. A vehicle has an overheating concern. Technician A says that the water pump could be intermittently working. Technician B says that the thermostat could be intermittently sticking. Who is correct?

    A. A only
    B. B only
    C. Both A and B
    D. Neither A nor B

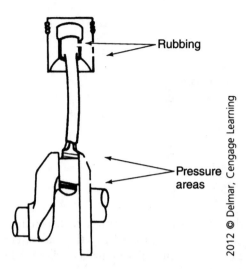

25. Referring to the above illustration, this condition may cause all of the following EXCEPT:

    A. Uneven connecting-rod bearing wear
    B. Uneven main bearing wear
    C. Uneven piston pin wear
    D. Uneven cylinder wall wear

26. A low-pitched roaring noise is being diagnosed. The noise disappears when the accessory drive belt is removed. Technician A says the belt tensioner pulley bearing should be checked for roughness. Technician B says the alternator bearing should be checked for roughness. Who is correct?

    A. A only
    B. B only
    C. Both A and B
    D. Neither A nor B

27. A truck has set a DTC PO171 (system too lean, bank 1). No other drivability concerns are present. The freeze frame data shows the code was set under warm idle conditions. Technician A says the problem could be an intake manifold vacuum leak. Technician B says the problem could be a weak fuel pump. Who is correct?

    A. A only
    B. B only
    C. Both A and B
    D. Neither A nor B

28. The LEAST LIKELY cause of spark knock is:

    A. EGR valve stuck closed
    B. Fuel quality
    C. Carbon buildup on top of the pistons
    D. EGR valve stuck open

29. Technician A says the engine should be disabled to prevent engine start up during a starter current draw test. Technician B says that on some vehicles, the accelerator pedal can be held to the floor to prevent starting during a current draw test. Who is correct?

    A. A only
    B. B only
    C. Both A and B
    D. Neither A nor B

30. Why are some fuel pressure regulators vacuum operated?

    A. To increase fuel delivery under high load conditions
    B. To prevent fuel-pressure leak down when the engine is turned off
    C. To provide a constant pressure drop across the injector due to changing manifold pressure
    D. To improve injector spray patterns

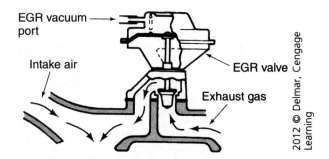

31. In the above illustration, Technician A says that EGR valves with heavy carbon build-up must be cleaned with a rotary file. Technician B says with the valve off the engine, fuel-injection cleaner may be used to clean the EGR valve's internal exhaust passages. Who is correct?

    A. A only
    B. B only
    C. Both A and B
    D. Neither A nor B

32. Technician A says that a valve with a margin measurement that is smaller than specified should be replaced. Technician B says that a valve with a stem measurement smaller than specified must be replaced. Who is correct?

    A. A only
    B. B only
    C. Both A and B
    D. Neither A nor B

33. Technician A says that enabling criteria are the specific conditions that must be met, such as ambient temperature or engine load, before a monitor will run. Technician B says that pending conditions are conditions that exist that prevent a specific monitor from running, such as an oxygen sensor fault code preventing a catalyst monitor from running. Who is correct?

    A. A only
    B. B only
    C. Both A and B
    D. Neither A nor B

34. A four-gas exhaust emissions analyzer may be used to help diagnose all of the following problems EXCEPT:

    A. A slow or lazy oxygen sensor
    B. Engine cylinder misfire
    C. A defective catalytic converter
    D. A plugged or restricted fuel injector

35. A vehicle stalls intermittently at idle and has low long-term fuel trim correction values stored when checked with a scan tool. All of the following conditions could cause this EXCEPT:

    A. Leaking fuel injectors
    B. Unmetered air leaking into the engine
    C. A leaking fuel-pressure regulator diaphragm
    D. A fuel-pressure regulator sticking closed

36. A truck comes into the shop with an oil pressure light flickering at idle. Technician A says the truck needs a new oil pump to correct the problem. Technician B says a known good oil pressure gauge should be installed on the engine to properly verify oil pressure. Who is correct?

    A. A only
    B. B only
    C. Both A and B
    D. Neither A nor B

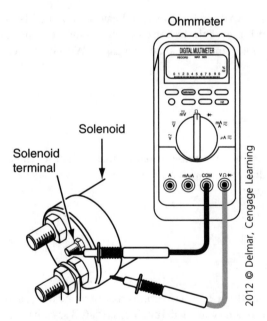

37. What procedure is being performed in the above illustration?

    A. Starter solenoid resistance test
    B. Ignition current draw test
    C. Battery load test
    D. Starter solenoid available voltage test

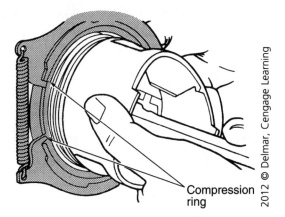

Compression
ring

2012 © Delmar, Cengage Learning

38. The tool in the above illustration is being used for what purpose?

   A. To widen the piston ring grooves
   B. To deepen the piston ring grooves
   C. To remove and replace the piston rings
   D. To remove carbon from the piston ring grooves

39. An engine equipped with a distributorless ignition system (DIS) will not start. Technician A says a defective CKP could cause this. Technician B says an open ground wire to the DIS assembly could be the cause. Who is correct?

   A. A only
   B. B only
   C. Both A and B
   D. Neither A nor B

40. Which of the following would most likely cause a single cylinder misfire?

   A. 5 percent cylinder leakage
   B. 50 percent cylinder Leakage
   C. A shorted CKP
   D. Retarded ignition timing

41. Technician A says that low cranking speed can be caused by high electrical resistance in the starter electrical circuit. Technician B says that a spun rod bearing can be the cause of the starter motor not being able to turn the engine over. Who is correct?

   A. A only
   B. B only
   C. Both A and B
   D. Neither A nor B

42. Technician A says that an EGR position sensor may be tested with either a voltmeter or an oscilloscope. Technician B says that if the PCM does not set an EGR position sensor fault code, there is no problem with the sensor. Who is correct?

   A. A only
   B. B only
   C. Both A and B
   D. Neither A nor B

43. A primary ignition circuit on a vehicle checks good, but there is no spark from the coil wire. This could be caused by:

    A. A defective coil
    B. A grounded rotor
    C. An overheated transistor
    D. An open diode

44. Blue smoke emission from the tailpipe can be caused by all of the following EXCEPT:

    A. Worn valve guides
    B. Worn valve seals
    C. Fouled spark plugs
    D. Worn piston rings

45. Technician A says that if the piston ring end gap is less than specified, a large piston ring must be used. Technician B says less than specified piston ring end gap can cause a piston ring to break. Who is correct?

    A. A only
    B. B only
    C. Both A and B
    D. Neither A nor B

46. A truck with an electric fuel pump must be cranked for an excessive time before the engine will start. Technician A says a fuel leak down test should be performed. Technician B says there may be a fault in the fuel pump relay circuit. Who is correct?

    A. A only
    B. B only
    C. Both A and B
    D. Neither A nor B

47. When performing a cylinder leakage test, which of the following is usually considered to be the maximum acceptable percentage?

    A. 5 to 10 percent
    B. 15 to 20 percent
    C. 30 to 40 percent
    D. 40 to 45 percent

48. A truck is brought to the shop with a fuel-related DTC. Technician A says all codes should be recorded and then erased to see if any are active codes. Technician B says some scan tools list active and inactive for ease of interpretation. Who is correct?

    A. A only
    B. B only
    C. Both A and B
    D. Neither A nor B

49. Engine valve springs should be inspected for all of the following EXCEPT:

    A. Cracks
    B. Spring tension
    C. Temperature resistance
    D. Spring squareness

50. While performing a starter current draw test on a truck, a technician records 75 amps. Technician A says this indicates excessive resistance in the starter control circuit. Technician B says 75 amps is an excessive current draw. Who is correct?

    A. A only
    B. B only
    C. Both A and B
    D. Neither A nor B

PREPARATION EXAM 5

1. Technician A says valve adjustment should always be performed on a running engine. Technician B says the piston should be placed at top dead center (TDC) of the compression stroke. Who is correct?

    A. A only
    B. B only
    C. Both A and B
    D. Neither A nor B

2. The ignition control module uses the digital signal received from the powertrain control module (PCM) for:

    A. RPM input
    B. Hall effect timing
    C. Cylinder #1 signal
    D. Computed timing signal

3. Technician A says a faulty throttle position sensor (TPS) can cause a hesitation on acceleration. Technician B says a faulty TPS will always set a diagnostic trouble code (DTC). Who is correct?

    A. A only
    B. B only
    C. Both A and B
    D. Neither A nor B

4. The LEAST LIKELY cause of a fuel tank leak is:

   A. Defective tank straps
   B. Road damage
   C. Defective seams
   D. Corrosion

5. When replacing a programmable read only memory (PROM), Technician A says that you should never ground yourself to the vehicle. Technician B says that grounding yourself to the vehicle will erase the PROM. Who is correct?

   A. A only
   B. B only
   C. Both A and B
   D. Neither A nor B

6. A slipped timing belt can cause all of the following EXCEPT:

   A. Poor fuel mileage
   B. A no-start condition
   C. High manifold vacuum
   D. Low power

7. Technician A says a vacuum leak decreases engine performance. Technician B says propane is a good method of locating vacuum leaks. Who is correct?

   A. A only
   B. B only
   C. Both A and B
   D. Neither A nor B

8. A computer ground circuit is being tested for voltage drop. Which reading below would be acceptable?

   A. 0.05 V
   B. 0.9 V
   C. 1.1 V
   D. 1.9 V

9. A magnetic pulse crank sensor is being tested with an ohmmeter. Technician A says when the pickup coil leads are moved, an erratic ohmmeter reading is normal. Technician B says that an infinite ohmmeter reading between the pickup coil terminals is an acceptable reading. Who is correct?

   A. A only
   B. B only
   C. Both A and B
   D. Neither A nor B

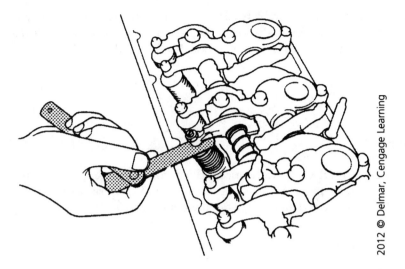

2012 © Delmar, Cengage Learning

10. Technician A says that if the measurement in the above illustration is set too wide, it will retard valve timing. Technician B says it will reduce valve overlap. Who is correct?

   A. A only
   B. B only
   C. Both A and B
   D. Neither A nor B

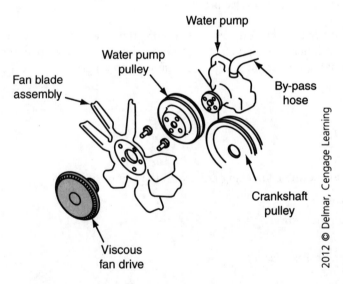

2012 © Delmar, Cengage Learning

11. A technician is testing a viscous-drive fan clutch, as shown in the above illustration, with the engine off. When the technician rotates the cooling fan by hand, it should have:

   A. More resistance hot
   B. More resistance cold
   C. No rotation movement
   D. No resistance

12. A vehicle is being diagnosed for a drivability complaint. The technician finds a soft code stored in the PCM. Technician A says the soft code is a code that would be classified as a C-type code. Technician B says the soft code is a code that occurred in the past and no longer exists. Who is correct?

    A. A only
    B. B only
    C. Both A and B
    D. Neither A nor B

13. Technician A says that with the PCV valve disconnected from the rocker cover, there should be no vacuum at the valve with the engine idling. Technician B says that when the PCV valve is removed and shaken, there should not be a rattling noise. Who is correct?

    A. A only
    B. B only
    C. Both A and B
    D. Neither A nor B

14. All of the following are true of the cylinder leakage test EXCEPT:

    A. Air loss and bubbles in the radiator indicate a bad head gasket or engine casting crack.
    B. Air loss from the oil filler cap indicates worn piston rings.
    C. A gauge reading of 100 percent indicates no cylinder leakage.
    D. Air loss from the exhaust indicates a valve problem.

15. Technician A says a key-off current draw on a battery of 0.05 amps is acceptable for most cars. Technician B says that under most operating conditions, the alternator supplies the current for vehicle electrical loads once the engine is running. Who is correct?

    A. A only
    B. B only
    C. Both A and B
    D. Neither A nor B

16. Which of these would cause a double knocking noise with the engine at an idle?

    A. Worn piston wrist pins
    B. Excessive timing chain deflection
    C. Worn main bearing
    D. Excessive main bearing thrust clearance

17. An engine has a lack of power and excessive fuel consumption. Technician A says a broken timing belt cannot be the cause. Technician B says the timing belt may have jumped a tooth. Who is correct?

    A. A only
    B. B only
    C. Both A and B
    D. Neither A nor B

18. When diagnosing fuel-injection system problems, a technical service bulletin search is performed for all of the following reasons EXCEPT:

    A. To save diagnostic time
    B. To locate service manual updates or specification changes
    C. Mid-year production changes
    D. Year, make, and model identification

19. A test lamp is connected between the negative side of the coil and ground to diagnose a no-start condition. Technician A says a flickering test lamp could be caused by a defective ignition module. Technician B says a flickering test lamp could be caused by a defective pickup coil. Who is correct?

    A. A only
    B. B only
    C. Both A and B
    D. Neither A nor B

20. All of the following are part of PCM reprogramming EXCEPT:

    A. Connect battery maintainer.
    B. Check technical service bulletins (TSBs) for current updates.
    C. Disconnect the negative battery cable.
    D. Turn ignition key to the on position.

21. Technician A says a fuel-pressure test will test the ability of the fuel pump to provide pressure. Technician B says it is possible for a fuel pump to pass a fuel-pressure test and still fail to provide sufficient flow. Who is correct?

    A. A only
    B. B only
    C. Both A and B
    D. Neither A nor B

22. A vehicle has a rough idle with black smoke coming out of the exhaust. The engine smoothes out when the exhaust gas recirculation (EGR) valve is lightly tapped with a hammer. Technician A says the EGR valve vacuum hose could be leaking. Technician B says the EGR valve pressure differential sensor could be faulty. Who is correct?

    A. A only
    B. B only
    C. Both A and B
    D. Neither A nor B

23. The LEAST LIKELY cause of blue exhaust smoke is:

    A. Worn valve seats
    B. Worn valve seals
    C. Stuck piston rings
    D. Worn cylinder walls

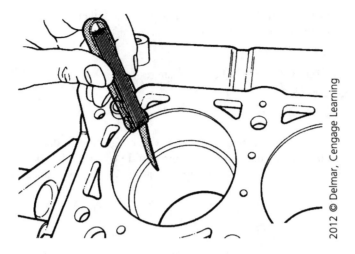

2012 © Delmar, Cengage Learning

24. When measuring the piston ring end gap, as shown in the above illustration, Technician A says that the ring gap should be measured with the ring positioned at the middle of the ring travel in the cylinder. Technician B says that ring gap should not exceed 0.004 inches in a 4-inch diameter bore. Who is correct?

    A. A only
    B. B only
    C. Both A and B
    D. Neither A nor B

25. Technician A says the ignition module may control ignition coil dwell time on some ignition systems. Technician B says the ignition module may control ignition coil current flow on some ignition systems. Who is correct?

    A. A only
    B. B only
    C. Both A and B
    D. Neither A nor B

26. While monitoring secondary ignition with an oscilloscope, the LEAST LIKELY cause of high resistance in the ignition secondary circuit is:

    A. Damaged spark plug wires
    B. No dielectric compound on the ignition module mounting surface
    C. Corroded spark plug wire ends
    D. Excessive spark plug air gap

27. A vehicle with a MIL lamp illuminated is being diagnosed. The DTC stored in the PCM is PO303 (cylinder #3 misfire). Technician A says this is a one-trip failure because catalyst damage can occur. Technician B says the MIL will flash when this code is stored.
    Who is correct?

    A. A only
    B. B only
    C. Both A and B
    D. Neither A nor B

28. A vacuum gauge is connected to an engine. When the engine is accelerated and held at a steady speed, the gauge slowly drops to 8″ vacuum. Which of the following could be the problem?

    A. Sticking valves
    B. Over-advanced ignition timing
    C. A restricted exhaust
    D. A rich fuel mixture

29. Technician A says the electrolyte level is not important in a non-serviceable battery. Technician B says that on some batteries, the electrolyte level can be checked in a sealed battery by looking through the translucent battery case. Who is correct?

    A. A only
    B. B only
    C. Both A and B
    D. Neither A nor B

30. Technician A says some evaporative emissions canisters have a replaceable filter. Technician B says if the filler cap is equipped with pressure and vacuum valves, they must be checked for dirt contamination and damage. Who is correct?

    A. A only
    B. B only
    C. Both A and B
    D. Neither A nor B

31. A technician is performing a compression test. Which statement below is LEAST LIKELY to be true?

    A. Higher than normal readings on all cylinders could be caused by carbon buildup.
    B. Even but lower than normal readings on all cylinders could be caused by a slipped timing chain.
    C. Low readings on two adjacent cylinders might be caused by a blown head gasket.
    D. A low reading on one cylinder may be caused by a vacuum leak at that cylinder.

32. A vehicle needs to have its PCM re-flashed. Technician A says the vehicle should be connected to a battery maintainer to prevent accidental battery discharge during flashing. Technician B says the scan tool connector must fit snugly in the diagnostic link connector (DLC). Who is correct?

    A. A only
    B. B only
    C. Both A and B
    D. Neither A nor B

33. While performing a valve adjustment, Technician A says the crankshaft must be placed in the proper position so that the valve lifter is riding on the camshaft lobe. Technician B says that adjusting valves with too little clearance may cause rough running and burnt valves. Who is correct?

    A. A only
    B. B only
    C. Both A and B
    D. Neither A nor B

34. While testing the cooling system, Technician A says to repeat the pressure test after repairs are made to ensure that all leaks are found. Technician B says a pressure test should include testing the radiator cap. Who is correct?

    A. A only
    B. B only
    C. Both A and B
    D. Neither A nor B

35. The LEAST LIKELY cause of low cylinder compression is:

    A. Worn valves
    B. Worn rings
    C. Blown head gasket
    D. Worn valve guides

36. Technician A says that during a PCM replacement, the technician should use a grounding strap to prevent static charge from damaging the PCM. Technician B says care should be taken not to touch the PCM terminals with your fingers. Who is correct?

    A. A only
    B. B only
    C. Both A and B
    D. Neither A nor B

37. Technician A says that a nylon fuel line that is bent sharply, causing a kink, is allowable and will not affect fuel flow. Technician B says nylon fuel line cannot be repaired, and the entire line must be replaced. Who is correct?

    A. A only
    B. B only
    C. Both A and B
    D. Neither A nor B

38. During a cylinder power balance test, there is no RPM drop on cylinder #3. Technician A says that the cylinder may have an inoperative injector. Technician B says that the cylinder may have an inoperative spark plug. Who is correct?

    A. A only
    B. B only
    C. Both A and B
    D. Neither A nor B

39. The LEAST LIKELY symptom resulting from an evaporative emissions (EVAP) system failure is:

    A. Increased tailpipe emissions
    B. Low vehicle emissions
    C. Malfunction indicator lamp (MIL) illumination
    D. Fuel odor

40. A pulse-style, secondary air-injection system is driven:

    A. By a drive belt
    B. Hydraulically
    C. By an electric motor
    D. With negative pressure pulses in the exhaust system

41. A truck that has low power also has low intake manifold vacuum. Technician A says a restricted air filter could be the cause. Technician B says a restricted exhaust could be the cause. Who is correct?

    A. A only
    B. B only
    C. Both A and B
    D. Neither A nor B

42. Technician A says a thorough ignition coil test includes both primary and secondary winding resistance tests. Technician B says maximum coil output testing can be performed with an oscilloscope. Who is correct?

    A. A only
    B. B only
    C. Both A and B
    D. Neither A nor B

43. Technician A says the filter on the engine off, natural vacuum (EONV) EVAP system should be serviced at regular oil change intervals. Technician B says the EONV EVAP system does not use a vacuum pump. Who is correct?

    A. A only
    B. B only
    C. Both A and B
    D. Neither A nor B

44. The leak down detection system is designed to:

    A. Pressurize the EVAP system to check for leaks
    B. Control the idle speed
    C. Transfer fuel from the tank to the injectors
    D. Stabilize engine RPM under a load

45. During a cylinder leak down test on a 4-cylinder engine, air is heard coming from spark plug hole #3 as cylinder #4 is being checked. Technician A says that this could be caused by a blown head gasket. Technician B says this could be caused by a cracked engine block. Who is correct?

    A. A only
    B. B only
    C. Both A and B
    D. Neither A nor B

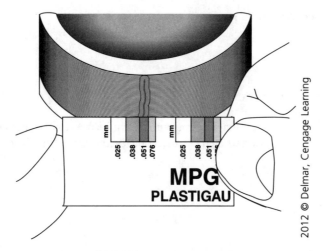

2012 © Delmar, Cengage Learning

46. In the above illustration, which of the following would be true?

  A. Crank end-play is 0.051 mm.
  B. Bearing crush is 0.051 mm.
  C. Bearing clearance is 0.051 mm.
  D. Crank runout is 0.051 mm.

47. An engine is making a loud metallic knocking that gets louder as the engine warms up or if the throttle is quickly snapped open. The noise almost disappears when the spark for cylinder #3 is shorted to ground. Technician A says the symptoms indicate a cracked flywheel. Technician B says the symptoms indicate a loose connecting-rod bearing. Who is correct?

  A. A only
  B. B only
  C. Both A and B
  D. Neither A nor B

48. A vehicle with an electronic ignition fails to start. Technician A says this could be caused by a defective crankshaft sensor connection. Technician B says this could be caused by a defective ignition module. Who is correct?

  A. A only
  B. B only
  C. Both A and B
  D. Neither A nor B

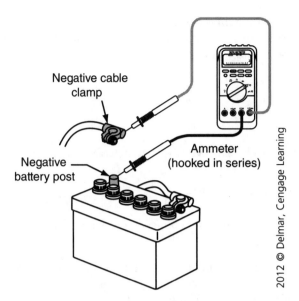

Negative cable clamp

Negative battery post

Ammeter (hooked in series)

2012 © Delmar, Cengage Learning

49. An ammeter, set in the milliamp position, is connected in series between the negative battery cable and ground, as shown in the above illustration. What is being measured?

   A.   Starter draw
   B.   Battery drain
   C.   Regulated voltage
   D.   Voltage drop

50. A vehicle with a no-start condition is being diagnosed. The vehicle has 12 volts at the fuel pump connector while cranking, but the fuel pump does not run. Technician A says the fuel pump could be the cause. Technician B says the ground side of the fuel pump circuit should be checked for open or high resistance. Who is correct?

   A.   A only
   B.   B only
   C.   Both A and B
   D.   Neither A nor B

PREPARATION EXAM 6

   1. A technician is performing a running compression test on a vehicle with suspected cylinder sealing problems. Technician A says running compression should be half of static compression at idle. Technician B says during a running compression test, the technician should snap the throttle, and the running compression should be 80 percent of static compression. Who is correct?

   A.   A only
   B.   B only
   C.   Both A and B
   D.   Neither A nor B

2. Technician A says the secondary ignition system is designed to handle voltages as high as 90,000 volts. Technician B says low secondary system voltage output could be the result of high primary ignition circuit resistance. Who is correct?

   A. A only
   B. B only
   C. Both A and B
   D. Neither A nor B

3. Which of the following steps is the technician LEAST LIKELY to perform when pressing the wrist pin into the piston and connecting rods?

   A. Align the bores in the piston and connecting rod.
   B. Heat the small end of the rod.
   C. Make sure that position marks on the piston and connecting rod are oriented properly.
   D. Heat the wrist pin.

4. Technician A says some vehicles require that the minimum airflow be adjusted at a throttle body service. Technician B says that some vehicles require that the minimum airflow be adjusted at the manifold absolute pressure sensor. Who is correct?

   A. A only
   B. B only
   C. Both A and B
   D. Neither A nor B

5. A truck has an occasional stumble on acceleration. There are no diagnostic trouble codes (DTCs) set. Which of the following could be the cause?

   A. Restricted exhaust
   B. Restricted air filter
   C. Open coolant temperature sensor
   D. Faulty APP sensor

6. Technician A says the exhaust gas recirculation (EGR) system is used to lower combustion chamber temperature. Technician B says that EGR systems that use an EGR valve position sensor should read about 4.5 volts with the EGR valve at full open. Who is correct?

   A. A only
   B. B only
   C. Both A and B
   D. Neither A nor B

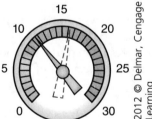

7. An engine is idling at 750 rpm. The pointer on the vacuum gauge in the above illustration is floating between 11 and 16 in. Hg. The most likely cause would be:

   A. Retarded timing
   B. Advanced timing
   C. A stuck EGR system valve
   D. Too lean an idle mixture

8. An engine equipped with electronic fuel injection has a loose exhaust manifold. Technician A says that the loose manifold may cause poor drivability. Technician B says that the loose manifold may cause an oxygen sensor code to be set. Who is correct?

    A. A only
    B. B only
    C. Both A and B
    D. Neither A nor B

9. A vehicle with DTC PO302 cylinder #2 misfire is being diagnosed. Technician A says low fuel pump pressure could be the cause. Technician B says a fouled spark plug could be the cause. Who is correct?

    A. A only
    B. B only
    C. Both A and B
    D. Neither A nor B

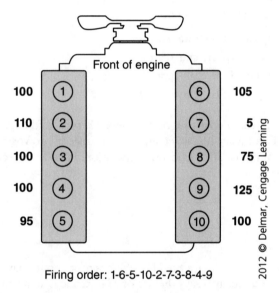

Firing order: 1-6-5-10-2-7-3-8-4-9

10. Cylinder power balance test results from a fuel-injected engine with a coil-on-plug ignition are shown in the above illustration. One cylinder is found to have virtually no RPM change. Which of these is the most likely cause?

    A. A faulty fuel injector
    B. A faulty crank position sensor
    C. A vacuum leak at the throttle body
    D. A fuel saturated vapor canister

11. A vehicle is being diagnosed for a rough idle. The vehicle exhaust has no visible color; however, the exhaust has a very strong odor and burns the eyes. Technician A says the vehicle could have a stuck closed pressure regulator causing an excessively lean mixture. Technician B says the vehicle could have a restricted fuel filter causing it to run too lean. Who is correct?

    A. A only
    B. B only
    C. Both A and B
    D. Neither A nor B

12. Technician A says a special puller and installer tool may be required to remove and install the vibration damper. Technician B says if the inertia ring on the vibration damper is loose, the damper must be replaced. Who is correct?

    A. A only
    B. B only
    C. Both A and B
    D. Neither A nor B

13. A truck has a short-term trim value that indicates the PCM is adding fuel. There is also a strong sulfur smell in the vehicle's exhaust. Which of the following is the most likely cause?

    A. A lean fuel mixture
    B. Coolant leaking into a combustion chamber
    C. A rich fuel mixture
    D. A vacuum leak

14. An engine that has a single cylinder misfire is being diagnosed by performing a leak down test. Which of these is the most likely cause of the condition?

    A. 45 percent leakage with air coming out of the intake valve
    B. 10 percent leakage with air coming out of the crankcase
    C. 5 percent leakage with air coming out of the crankcase
    D. 10 percent leakage with air coming out of the exhaust valve

15. Technician A says that secondary air is injected upstream of the converter on a cold engine to aid in heating up the oxygen sensor. Technician B says secondary air is injected downstream to the converter on a warm engine to aid in catalytic converter operation. Who is correct?

    A. A only
    B. B only
    C. Both A and B
    D. Neither A nor B

16. A Hall effect sensor is being tested. Technician A says the Hall effect sensor should have a resistance value of 500 to 800 ohms. Technician B says the use of a lab scope is an accurate method of checking Hall effect sensor operation. Who is correct?

    A. A only
    B. B only
    C. Both A and B
    D. Neither A nor B

17. There is a slight vapor from the exhaust accompanied by a sweet odor. Which of the following could be the cause?

    A. A leaking exhaust manifold
    B. A leaking intake manifold
    C. A faulty barometric pressure (BARO) sensor
    D. A faulty accelerator pedal position (APP) sensor

18. A truck with an under-vehicle rattle and low engine power is being diagnosed. Technician A says an intake manifold vacuum test could be performed to help locate a restricted exhaust. Technician B says the catalytic converter may have come apart and may be restricting exhaust flow. Who is correct?

    A.   A only
    B.   B only
    C.   Both A and B
    D.   Neither A nor B

2012 © Delmar, Cengage Learning

19. In the illustration above, the technician is most likely checking:

    A.   Valve guide depth
    B.   Valve seat angle
    C.   Cylinder head flatness
    D.   Valve seat runout

20. A vehicle is being diagnosed for an overheating complaint while at cruising speeds. Technician A says the electric cooling fan may be defective. Technician B says that the radiator may have some damaged cooling fins. Who is correct?

    A.   A only
    B.   B only
    C.   Both A and B
    D.   Neither A nor B

21. Oil is leaking from an engine's crankshaft rear main bearing seal. Technician A says the oil seal could be faulty. Technician B says the positive crankcase ventilation (PCV) system may not be functioning. Who is correct?

    A.   A only
    B.   B only
    C.   Both A and B
    D.   Neither A nor B

22. If new rings are installed without removing the ring ridge, which of these is the most likely result?

    A. Piston skirt damage
    B. Piston pin damage
    C. Connecting-rod bearing damage
    D. Piston compression ring damage

23. Technician A says most 4-wire COP (coil-on-plug) ignition systems use individual ignition control modules for each cylinder. Technician B says a two-wire COP coil is tested like any other ignition coil. Who is correct?

    A. A only
    B. B only
    C. Both A and B
    D. Neither A nor B

24. The customer is concerned by a loud popping noise, most noticeable during acceleration. Which of these is the most likely cause?

    A. An intake manifold leak
    B. A cooling system leak
    C. A missing air filter
    D. An exhaust manifold leak

25. Technician A says the powertrain control module (PCM) controls the amount of canister purge. Technician B says the evaporative emissions (EVAP) system controls the amount of NO_X emissions produced by the engine. Who is correct?

    A. A only
    B. B only
    C. Both A and B
    D. Neither A nor B

26. The maximum secondary voltage from an ignition coil is lower than specification. Which of the following could be the cause?

    A. An open primary winding
    B. An open secondary winding
    C. High resistance in the primary winding
    D. High resistance in the secondary ignition circuit

27. A vehicle with a rich exhaust code is being diagnosed. Technician A says the fuel pressure could be the cause. Technician B says a ruptured fuel-pressure regulator diaphragm could be the cause. Who is correct?

    A. A only
    B. B only
    C. Both A and B
    D. Neither A nor B

28. After a vehicle is parked overnight and then started in the morning, the engine has a lifter noise that disappears after it has been running for a short while. The most likely cause would be:

    A. Low oil pressure
    B. Low oil level
    C. A worn lifter bottom
    D. Excessive lifter leak down

29. Technician A says the PCV valve flow volume is low at wide-open throttle (WOT). Technician B says the PCV valve flow volume is low during deceleration. Who is correct?

    A. A only
    B. B only
    C. Both A and B
    D. Neither A nor B

30. Technician A says valve stem height is measured from the bottom of the valve guide to the top of the valve guide. Technician B says that excessive valve stem-to-guide clearance may result in excessive oil consumption. Who is correct?

    A. A only
    B. B only
    C. Both A and B
    D. Neither A nor B

2012 © Delmar, Cengage Learning

31. When using a compression tester, as shown in the above illustration, the compression readings on the cylinders are all even, but lower than the specified compression. This could indicate:

    A. A blown head gasket
    B. Carbon buildup
    C. A cracked head
    D. Worn rings and cylinders

32. On a truck equipped with coil-on-plug ignition, one primary circuit current flow is higher than specification. Which of the following could be the cause?

    A. High primary coil resistance
    B. High secondary coil resistance
    C. Low secondary coil resistance
    D. Low primary coil resistance

33. A powertrain control module is being replaced. Technician A says the new PCM should be ordered using the original PCM's part number. Technician B says installing a used PCM may cause the theft deterrent system to activate on some vehicles. Who is correct?

    A. A only
    B. B only
    C. Both A and B
    D. Neither A nor B

34. A cylinder power balance test is being performed on an engine to determine which cylinder is causing a misfire. Technician A says that when the faulty cylinder is disabled, engine RPM will drop more than for the other cylinders. Technician B says that disabling the faulty cylinder will cause the engine to stall. Who is correct?

    A. A only
    B. B only
    C. Both A and B
    D. Neither A nor B

35. The customer says that the engine requires excessive cranking to start. The LEAST LIKELY cause of this problem would be:

    A. A cracked cylinder block
    B. A jumped timing belt
    C. A faulty fuel pump
    D. A stuck open EGR valve

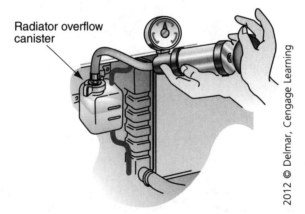

Radiator overflow canister

2012 © Delmar, Cengage Learning

36. The tester in the illustration above may be used to test all the following items EXCEPT:

    A. Coolant leaks
    B. The radiator cap pressure relief valve
    C. Freeze protection level
    D. Heater core leaks

37. A throttle position sensor (TPS) is being tested with a digital multi-meter (DMM). Technician A says the voltage on the signal wire should be around 1.0 volt or less at idle. Technician B says the voltage reading should be watched for irregularities while sweeping the TPS. Who is correct?

    A. A only
    B. B only
    C. Both A and B
    D. Neither A nor B

38. A voltage drop test is being performed on an ignition coil power supply wire. When performing this test:

    A. The key should be off.
    B. The engine should be running.
    C. The key should be on, but the engine should not be running.
    D. The vehicle should be driven above 35 mph.

39. A vehicle is being diagnosed for a PO134 DTC (oxygen sensor circuit, no activity detected on bank 1, sensor 1). Technician A says the problem could be low fuel system pressure. Technician B says the problem could be the downstream oxygen sensor on the bank where the #1 cylinder is located. Who is correct?

    A. A only
    B. B only
    C. Both A and B
    D. Neither A nor B

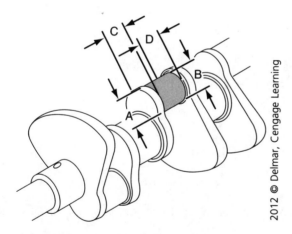

2012 © Delmar, Cengage Learning

40. When measuring the crankshaft journal, as shown in the above illustration, the difference between measurements:

    A. A and B indicates out of round.
    B. C and D indicates out of round.
    C. A and C indicates out of round.
    D. A and D indicates taper.

41. Technician A says group-fired injection systems fire more than one injector at the same time. Technician B says sequential fuel injection systems fire two or more injectors at the same time. Who is correct?

    A. A only
    B. B only
    C. Both A and B
    D. Neither A nor B

42. On an overhead camshaft (OHC) pressure cylinder head with removable bearing caps, which of the following is used to measure bearing alignment?

    A. A straightedge and a feeler gauge
    B. Plastigauge
    C. A dial indicator
    D. A telescoping gauge

43. Air is heard escaping from the exhaust during a cylinder leakage test. Technician A says this condition could indicate a leaking exhaust valve. Technician B says the technician must ensure that the cylinder is at top dead center (TDC) of the compression stroke when performing this test. Who is correct?

    A. A only
    B. B only
    C. Both A and B
    D. Neither A nor B

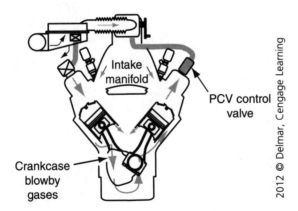

44. The above illustration shows the PCV system. All of the following are symptoms of a stuck open PCV valve EXCEPT:

    A. High manifold vacuum
    B. Engine stalling
    C. Rough idle operation
    D. Lean air/fuel ratio

45. Which of these should be performed first when a starter fails to crank?

   A. Measure battery voltage.
   B. Remove spark plugs.
   C. Check for the presence of spark.
   D. By-pass the starter solenoid with the remote starter button.

46. Technician A says that measuring voltage drop can only be performed on a circuit that has current flow. Technician B says a high voltage drop reading indicates excessive resistance in the circuit. Who is correct?

   A. A only
   B. B only
   C. Both A and B
   D. Neither A nor B

47. A loose belt may cause all of these EXCEPT:

   A. A discharged battery
   B. Water pump bearing failure
   C. Poor power steering assist
   D. Engine overheating

48. Engine oil is found to be leaking from many different locations on an engine. Technician A says the PCV valve may be restricted. Technician B says the ignition system may be misfiring. Who is correct?

   A. A only
   B. B only
   C. Both A and B
   D. Neither A nor B

49. When measured on the supply line to the fuel rail, fuel pump flow volume is less than specified. Which of the following could be the cause?

   A. Restricted fuel injectors
   B. An electrically open fuel injector
   C. Restricted return line
   D. Low voltage to the fuel pump

50. Technician A says blue-gray smoke coming from the exhaust may be caused by stuck piston rings. Technician B says this could be caused by a plugged oil drain passage in the cylinder head. Who is correct?

   A. A only
   B. B only
   C. Both A and B
   D. Neither A nor B

Answer Keys and Explanations

INTRODUCTION

Included in this section are the answer keys for each preparation exam, followed by individual, detailed answer explanations and a reference identifying the designated task area being assessed by each specific question. This additional reference information may prove useful if you need to refer back to the task list located in Section 4 of this book for additional support.

PREPARATION EXAM 1—ANSWER KEY

1.	B	21.	D	41.	A
2.	D	22.	A	42.	C
3.	C	23.	D	43.	D
4.	B	24.	D	44.	D
5.	C	25.	B	45.	C
6.	A	26.	D	46.	C
7.	D	27.	C	47.	B
8.	B	28.	D	48.	D
9.	C	29.	C	49.	C
10.	A	30.	A	50.	D
11.	B	31.	D		
12.	A	32.	B		
13.	C	33.	B		
14.	B	34.	D		
15.	C	35.	A		
16.	C	36.	C		
17.	C	37.	C		
18.	D	38.	A		
19.	B	39.	C		
20.	B	40.	D		

PREPARATION EXAM 1—EXPLANATIONS

1. Which of the following would be the normal maximum amount of warpage allowed for a cast iron cylinder head?
    A. 0.0001″
    B. 0.003″
    C. 0.010″
    D. 0.030″

    TASK B.3, B.1

    Answer A is incorrect. 0.0001″ is an extremely small amount of warpage and would not be considered the normal maximum specification.

    Answer B is correct. 0.003″ is a normal maximum warpage specification for a cast iron cylinder head for a gasoline engine.

    Answer C is incorrect. 0.010″ is a substantial amount of warpage and would exceed normal specifications.

    Answer D is incorrect. 0.030″ is a substantial amount of warpage and would exceed normal specifications.

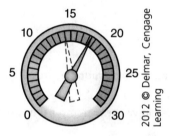

2012 © Delmar, Cengage Learning

2. With the engine idling, a vacuum gauge connected to the intake manifold fluctuates as shown in the above illustration. These vacuum gauge fluctuations may be caused by:
    A. Late ignition timing
    B. Intake manifold vacuum leak at the throttle body
    C. A restricted exhaust system
    D. Sticky valve stems and guides

    TASK A.6

    Answer A is incorrect. Late ignition timing would result in a low, steady reading.

    Answer B is incorrect. An intake manifold leak at the throttle body would cause a very low, steady reading.

    Answer C is incorrect. A restricted exhaust system would cause vacuum to slowly decrease.

    Answer D is correct. If the valve sticks in its guide, it may not completely close. This would result in compression loss and pulsing vacuum gauge.

3. Which of the following is the most likely cause of low oil pressure at idle?
    A. Worn piston rings
    B. Worn piston pins
    C. Worn bearings
    D. Worn cam lobes

    TASK D.1

    Answer A is incorrect. Worn piston rings cause low compression.

    Answer B is incorrect. Worn piston pins cause engine knocks.

    Answer C is correct. Worn bearings cause increased bearing clearances and decreased oil pressures.

    Answer D is incorrect. Worn cam lobes cause poor cylinder breathing and misfires.

TASK E.5

4. Which of the following tools would be used to bench check an ignition coil?

    A.   Voltmeter
    B.   Ohmmeter
    C.   Ammeter
    D.   Oscilloscope

    Answer A is incorrect. A voltmeter can be used for on-the-vehicle testing but would not be useful for bench testing.

    Answer B is correct. An ohmmeter can be used to measure the resistance of the windings in an ignition coil.

    Answer C is incorrect. An ammeter requires the ignition coil to be connected to the circuit.

    Answer D is incorrect. An oscilloscope requires the ignition coil to be connected to the circuit.

TASK F.1

5. A vehicle has low fuel pressure (5 psi) and will not start. Which of the following could be the cause?

    A.   An open fuel pump relay
    B.   An open fuel pump winding
    C.   Low fuel level
    D.   Low oil level

    Answer A is incorrect. An open relay would prevent the pump from running, and there would be no fuel pressure.

    Answer B is incorrect. An open fuel pump winding would prevent the pump from running, and there would be no fuel pressure.

    Answer C is correct. Low fuel level in the tank can cause low fuel pressure and prevent the engine from starting.

    Answer D is incorrect. Low engine oil level is not desirable and should be corrected; however, it would not cause low fuel pump pressure.

TASK A.1

6. The last step on most diagnostic flow charts is:

    A.   Verify the repair
    B.   Perform the repair
    C.   Identify the cause for concern
    D.   Verify concern

    Answer A is correct. Verifying the repair is the last step and would be done after the repair is complete.

    Answer B is incorrect. Performing the repair happens after the problem is identified, but it cannot be the last step. The repair must be verified after it is performed.

    Answer C is incorrect. Identifying the cause for the concern is one of the first steps, after verifying the concern. Repairs cannot happen until the cause is identified.

    Answer D is incorrect. The first step is usually to verify the cause of the concern.

7. The PCV valve is removed with the engine running. There is no change in engine operation. This would indicate:

TASK G.1

A. Normal engine operation

B. A plugged valve

C. A sticking valve

D. A restricted hose

Answer A is incorrect. The engine RPM should change if the system is operating properly.

Answer B is incorrect. If the valve is plugged, the engine RPM should still increase when the valve is removed.

Answer C is incorrect. If the valve is sticking, the engine RPM should still change when the valve is removed.

Answer D is correct. The hose is restricted. Removing the valve makes no difference in the amount of air entering the engine.

8. A surface comparator is used to:

TASK C.2

A. Compare surface warpage

B. Compare surface roughness

C. Compare head thickness

D. Compare block thickness

Answer A is incorrect. Surface warpage is measured with a straightedge and feeler gauge, not a surface comparator.

Answer B is correct. A surface comparator is a visual device used to compare the deck and head gasket surface roughness to manufacturers' specifications.

Answer C is incorrect. Cylinder head thickness is measured with an outside micrometer, not a surface comparator.

Answer D is incorrect. Comparing block thickness is not a task a technician would normally do.

9. All of the following is correct concerning diagnostic trouble codes (DTCs) EXCEPT:

TASK H.11

A. Some DTCs will not set until two complete drive cycles.

B. Some DTCs are erased when the battery is disconnected.

C. All DTCs are erased when the electronic control module (ECM) fuse is removed.

D. Some DTCs will set on the first drive cycle.

Answer A is incorrect. Some DTCs do require two complete drive cycles prior to setting the code.

Answer B is incorrect. Some DTCs require the use of a scan tool to erase them. However, on some vehicles, the DTCs can be erased by disconnecting the battery.

Answer C is correct. Some vehicles can have the DTCs removed by removing the ECM fuse. However, some vehicles require the use of a scan tool to erase them.

Answer D is incorrect. A possible catalyst-damaging event can set a DTC during the first drive cycle. DTCs, which do not affect emissions, will often take two trips to set a code.

10. Which of the following would most likely result from a restricted catalytic converter?

    A. Low power
    B. Increased engine idle RPM
    C. Increased engine vacuum
    D. Low combustion chamber temperatures

    Answer A is correct. A restricted catalytic converter would prevent the engine from breathing properly; this would result in low power.

    Answer B is incorrect. The engine idle RPM may decrease not increase, due to reduced airflow.

    Answer C is incorrect. The engine vacuum would decrease because the engine would not be able to breathe efficiently due to reduced airflow.

    Answer D is incorrect. A restricted catalytic converter would cause combustion chamber temperatures to increase because the exhaust gases would not readily escape. Low combustion chamber temperatures are often caused by a stuck open thermostat.

11. During a vehicle road test the vehicle blows blue smoke from the tailpipe. Which of the following could be the cause?

    A. Worn compression rings
    B. Worn oil control rings
    C. Worn intake valve seats
    D. Worn exhaust valve seats

    Answer A is incorrect. Worn compression rings will cause low compression.

    Answer B is correct. Worn oil control rings can allow oil into the combustion chamber and cause blue smoke.

    Answer C is incorrect. Worn intake vale seats would cause low compression.

    Answer D is incorrect. Worn exhaust valve seats would cause low compression

12. A truck with half a tank of gasoline has low power and stalls. The fuel tank is filled. The truck restarts, runs for an hour and stalls again. After refueling, the sequence starts again. Which of the following could be the cause?

    A. The fuel pump is failing.
    B. The fuel pump relay is failing.
    C. The ECM is failing.
    D. The ignition control module is failing.

    Answer A is correct. These symptoms are indicative of a fuel pump that is failing. The fuel being added to the tank helps to cool the failing pump.

    Answer B is incorrect. A failed fuel relay can cause a no-start condition, but adding fuel to the tank would not greatly affect the fuel pump relay.

    Answer C is incorrect. A failed ECM can cause a vehicle to intermittently stall, but adding fuel would not cause the engine to start and run normally again.

    Answer D is incorrect. A failing ignition module will sometimes allow the engine to restart after it cools, but the ignition control module would not be affected by refilling the fuel tank.

13. Cylinder head thickness is less than specified. The most likely cause is:

   A. Engine overheating
   B. Insufficient oil pressure
   C. Excessive milling
   D. Low engine operating temperature.

TASK B.3

Answer A is incorrect. Engine overheating can warp a cylinder head; however, it will not result in cylinder head thickness being less than specification.

Answer B is incorrect. Insufficient oil pressure can result in severe engine damage, but it will not affect cylinder head thickness.

Answer C is correct. A cylinder head that is thinner than specification is the result of excessive milling.

Answer D is incorrect. A low engine operating temperature can result in engine sludging; however, it would not cause the cylinder head to be thinner than specification.

14. The oil pressure gauge is reading lower than normal; what should the technician do first?

   A. Check the pressure with a master gauge
   B. Check the oil level
   C. Change the oil
   D. Change the oil filter

TASK D.1

Answer A is incorrect. When the dash gauge reads low, the integrity of the gauge and sending unit certainly should be questioned, but the first diagnostic step should be to check the oil level.

Answer B is correct. Low oil level can cause low oil pressure and is a common problem.

Answer C is incorrect. The oil should not be changed until it is determined necessary.

Answer D is incorrect. The oil filter should not be changed until it is determined necessary.

15. The acronym TSB means:

   A. Technical Sales Bulletin
   B. Trade School Books
   C. Technical Service Bulletin
   D. Trade Service Books

TASK A.2

Answer A is incorrect. TSB as used in the automotive field means Technical Service Bulletin. Sales bulletins are used by the sales staff.

Answer B is incorrect. TSB as used in the automotive field means Technical Service Bulletin, not Trade School Books.

Answer C is correct. TSB as used in the automotive field means Technical Service Bulletin. Manufacturers use technical service bulletins to update service technicians on the latest repair information.

Answer D is incorrect. TSB as used in the automotive field means Technical Service Bulletin, not Trade Service Books.

TASK G.6

16. Which tool can be used to check catalytic converter efficiency?

    A. Compression gauge
    B. Cylinder leakage tester
    C. Infrared pyrometer
    D. Refractometer

    Answer A is incorrect. A compression gauge can be used to locate a cylinder low on compression, but it would not be useful in isolating a failed converter.

    Answer B is incorrect. A cylinder leakage tester will help find a leaking intake valve, but it will not help in checking the converter.

    Answer C is correct. An infrared pyrometer can be used to measure the converter's inlet and outlet temperatures to determine if it is operating correctly.

    Answer D is incorrect. A refractometer is used to check coolant; it would not be helpful in checking a converter.

TASK C.3

17. All of the following can be used to measure a cylinder bore for wear EXCEPT:

    A. A dial bore gauge
    B. An inside micrometer
    C. An outside micrometer
    D. A telescoping gauge

    Answer A is incorrect. A dial bore gauge is a fast and accurate method to measure the cylinder bore.

    Answer B is incorrect. An inside micrometer is an effective method to measure a cylinder bore.

    Answer C is correct. An outside micrometer cannot be used by itself to measure a cylinder bore for wear.

    Answer D is incorrect. A telescoping gauge can be used to measure a cylinder bore for wear.

**TASK E.3,
A.11**

18. An engine has set a misfire code for cylinder #5. The technician swaps the ignition coil on cylinders 5 and 6. Now there is a misfire code for cylinder #6. This most likely indicates a:

    A. Failed ECM
    B. Failed primary wiring harness
    C. Low compression on cylinder #5
    D. Failed ignition coil

    Answer A is incorrect. If the ECM had been at fault, the DTC would not have moved with the ignition coil.

    Answer B is incorrect. If the primary wiring harness had been at fault, the DTC would not have changed.

    Answer C is incorrect. If compression had been low on cylinder #5, the DTC would not have moved to cylinder 6.

    Answer D is correct. The DTC moved with the ignition coil. The most likely cause of the fault is the ignition coil.

19. Which of the following can be used to help indicate the source of an engine oil leak?

    A. Green light
    B. Blacklight
    C. Yellow light
    D. Strobe light

    TASK A.2

    Answer A is incorrect. A green light is not used for leak detection. A blacklight is used with fluorescent dye to help locate leaks.

    Answer B is correct. A blacklight is used in conjunction with fluorescent dye to help locate engine oil leaks.

    Answer C is incorrect. A yellow light would not be useful in detecting an engine oil leak.

    Answer D is incorrect. A strobe light is not used to locate engine oil leaks. A strobe light is used to check ignition timing on some engines.

20. A fuel pump has a higher than normal current draw. Which of the following could be the cause?

    A. High resistance in the fuel pump relay
    B. Tight bushings in the fuel pump
    C. High resistance in the fuel pump wiring
    D. An open in the fuel pump relay

    TASK F.1

    Answer A is incorrect. High resistance in the fuel pump relay will cause the fuel pump amperage to be lower than normal.

    Answer B is correct. A tight bushing in the fuel pump can cause the pump to have a higher than normal current draw. This can result in eventual failure of the pump. The relay contacts will also be damaged due to the increased current draw.

    Answer C is incorrect. High resistance in the fuel pump wiring would cause lower than normal current draw.

    Answer D is incorrect. An open would result in no current draw.

21. Which of the following would be used to measure valve guide inside diameter?

    A. Inside micrometer
    B. Depth micrometer
    C. Dial indicator
    D. Split ball gauge

    TASK B.4

    Answer A is incorrect. An inside micrometer cannot measure this small of a diameter. It would be used to measure cylinder liner inside diameter.

    Answer B is incorrect. A depth micrometer can measure liner counterbore depth; it cannot measure inside diameter.

    Answer C is incorrect. A dial indicator is used to measure relative movement, not the inside of a small diameter.

    Answer D is correct. A split ball gauge, also known as a small hole gauge, can be used to measure a small inside diameter such as a valve guide inside diameter.

TASK A.6

22. When measuring the intake manifold vacuum at idle, the technician observes a regular downward pulsation, dropping from 17 in. Hg to 10 in. Hg, then immediately recovering to 17 in. Hg. Which of the following could be the cause?

    A. A leaking intake valve
    B. Worn piston rings
    C. Worn turbocharger bearings
    D. A leaking intake manifold

    Answer A is correct. A leaking intake valve could cause a downward pulse of the needle. The valve would allow pressure into the intake manifold during the compression and exhaust strokes, which would cause the intake manifold to have a regular vacuum fluctuation.

    Answer B is incorrect. Worn piston rings tend to cause a low steady vacuum reading.

    Answer C is incorrect. Worn turbocharger bearings can cause the impeller to scrape the housing but would not likely cause a vacuum fluctuation.

    Answer D is incorrect. A leaking intake manifold will usually cause a coolant leak or a low steady vacuum reading.

TASK H.2, A.11

23. A DTC has been set for a shorted coolant temperature sensor. The scan tool is monitored while the sensor wiring harness is disconnected. The scan tool display does not change. Which of the following could be the cause?

    A. The sensor is shorted.
    B. The sensor is open.
    C. The wiring harness is open.
    D. The wiring harness is shorted.

    Answer A is incorrect. If the sensor were shorted, the scan tool should have changed when the sensor was disconnected.

    Answer B is incorrect. If the sensor were open, the DTC would have been for an open coolant temperature sensor circuit.

    Answer C is incorrect. If the harness were open, the DTC would have been for an open coolant temperature sensor circuit.

    Answer D is correct. The wiring harness is shorted. The display data did not change when the sensor was disconnected. The DTC did not change because the wiring harness between the ECM and the sensor is shorted. Disconnecting the sensor did nothing to change the short. The ECM does not recognize that the sensor was disconnected.

TASK E.4

24. Which of the following resistance readings would indicate an open spark plug wire?

    A. 0 ohms
    B. 100 ohms
    C. 5000 ohms
    D. O.L.

    Answer A is incorrect. 0 ohms would indicate a very low resistance.

    Answer B is incorrect. 100 ohms would indicate a low resistance.

    Answer C is incorrect. 5000 ohms may be a normal resistance, depending on the length and construction of the wire.

    Answer D is correct. O.L. indicates an open circuit.

25. Which of the following can be used to help indicate the source of an engine coolant leak?

    A. Green light
    B. Blacklight
    C. Yellow light
    D. Strobe light

TASK A.3

Answer A is incorrect. A green light is not used for leak detection. A blacklight is used with fluorescent dye to help locate leaks.

Answer B is correct. A blacklight is used in conjunction with fluorescent dye to help locate engine coolant leaks.

Answer C is incorrect. A yellow light would not be useful in detecting an engine coolant leak.

Answer D is incorrect. A strobe light is not used to locate engine coolant leaks. A strobe light is used to check ignition timing on some engines.

26. A vehicle is equipped with a negative temperature coefficient coolant temperature sensor. Technician A says temperature sensor resistance will increase as temperature increases. Technician B says temperature sensor resistance will decrease as temperature decreases. Who is correct?

    A. A only
    B. B only
    C. Both A and B
    D. Neither A nor B

TASK H.6

Answer A is incorrect. Resistance will not increase as temperature increases.

Answer B is incorrect. Resistance will not decrease as temperature decreases.

Answer C is incorrect. Neither Technician is correct.

Answer D is correct. Neither Technician is correct. With a negative temperature coefficient coolant temperature sensor, resistance increases as the temperature goes down and decreases when the temperature goes up.

27. Which of the following can be used to help indicate the source of an engine vacuum leak?

    A. Compression gauge
    B. Cylinder leakage tester
    C. Smoke machine
    D. Oscilloscope

TASK A.3

Answer A is incorrect. A compression gauge will show the condition of the rings and valves within the engine. However, it will not help identify the source of a vacuum leak.

Answer B is incorrect. A cylinder leakage tester can be useful in engine diagnosis; however, it would not help in locating the source of an engine vacuum leak.

Answer C is correct. A smoke machine can be used to apply smoke to the engine intake manifold. The technician can watch for where the smoke leaks out to help identify the source of an engine vacuum leak.

Answer D is incorrect. An oscilloscope would be used to locate problems within the ignition system. However, it would not be useful in locating an engine oil leak.

TASK G.1

28. An engine has a high idle concern. When the PCV hose is crimped shut, the engine RPM drops to normal. Which of the following is indicated?

    A. A restricted positive crankcase ventilation (PCV) system
    B. A stuck open idle air control valve
    C. A stuck closed idle air control valve
    D. A failed open PCV valve

    Answer A is incorrect. A restricted PCV system would not cause an increase in idle RPM. It could cause excessive crankcase pressure and external oil leaks.

    Answer B is incorrect. A stuck open idle air control system would cause a high idle; however, if this were the cause, the idle would not drop when the PCV hose is crimped.

    Answer C is incorrect. A stuck closed idle air control valve would not cause a high idle. It could cause a low idle due to reduced idle airflow.

    Answer D is correct. A failed open PCV system could allow too much airflow. When the system is crimped shut, the RPM would drop.

TASK A.3

29. Which of the following can be used to help indicate the source of an evaporative emissions (EVAP) system leak?

    A. Compression gauge
    B. Cylinder leakage tester
    C. Smoke machine
    D. Oscilloscope

    Answer A is incorrect. A compression gauge will show the condition of the rings and valves within the engine. However, it will not help identify the source of an EVAP system leak.

    Answer B is incorrect. A cylinder leakage tester can be useful in engine diagnosis; however, it would not help in locating the source of an EVAP system leak.

    Answer C is correct. A smoke machine can be used to apply smoke to the EVAP system manifold. The technician can watch for where the smoke leaks out to help identify the source of an EVAP system leak.

    Answer D is incorrect. An oscilloscope would be used to locate problems within the ignition system. However, it would not be useful in locating an EVAP system leak.

TASK C.4

30. Which of the following is the most common method for installing in-block camshaft bearings?

    A. Driven in
    B. Heated, then dropped in
    C. Cooled, then dropped in
    D. Cast in

    Answer A is correct. Camshaft bearings that are located in the block are typically driven in with a camshaft bearing driver.

    Answer B is incorrect. Camshaft bearings are not heated prior to installation.

    Answer C is incorrect. Camshaft bearings are not cooled prior to installation.

    Answer D is incorrect. Camshaft bearings used to be cast into the block many years ago; however, now all in-block camshaft bearings are driven in.

31. A truck has a single cylinder misfire. There is no spark at the spark plug. There is spark at the plug wire. Which of the following could be the cause of the misfire?

TASK E.4

   A.   Low fuel pressure
   B.   Low compression
   C.   Worn cam lobe
   D.   Failed spark plug

Answer A is incorrect. Low fuel pressure can cause low power; it would not prevent spark at the spark plug.

Answer B is incorrect. Low compression can cause a misfire, but it would not prevent spark at the spark plug.

Answer C is incorrect. A worn cam lobe can cause a misfire, but it would not prevent a spark at the spark plug.

Answer D is correct. A failed spark plug could cause a single cylinder misfire and could be the cause of no spark.

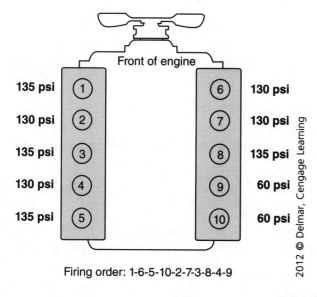

32. Compression test results for an engine are shown in the above illustration. Technician A says this could be caused by a loose timing chain. Technician B says a leaking head gasket could cause this. Who is correct?

TASK A.8

   A.   A only
   B.   B only
   C.   Both A and B
   D.   Neither A nor B

Answer A is incorrect. A loose timing chain would not affect just two cylinders.

Answer B is correct. Only Technician B is correct. When two adjacent cylinders have low compression, a leaking head gasket should be suspected.

Answer C is incorrect. Only Technician B is correct.

Answer D is incorrect. Technician B is correct.

TASK F.2

33. Short term fuel trim indicates the ECM is adding extra fuel. This could be caused by:

    A.   A stuck open fuel injector
    B.   A leaking exhaust manifold
    C.   A stuck closed fuel pressure regulator
    D.   A leaking exhaust tailpipe

    Answer A is incorrect. A stuck open fuel injector would add extra fuel, and the ECM would reduce fuel delivery in an attempt to compensate.

    Answer B is correct. A leaking exhaust manifold would allow extra oxygen into the exhaust, giving a false lean indication to the ECM.

    Answer C is incorrect. A stuck closed fuel pressure regulator would increase fuel pressure and delivery. The ECM would compensate by reducing fuel delivery.

    Answer D is incorrect. A leaking exhaust tailpipe would be post oxygen sensors; it would have no effect.

TASK B.4

34. Valve springs should be measured for all of the following EXCEPT:

    A.   Squareness
    B.   Free length
    C.   Tension
    D.   Radius

    Answer A is incorrect. Valve spring squareness is a vital check. An out of square spring can cause poor valve sealing.

    Answer B is incorrect. Free length is measured to determine if the spring has collapsed.

    Answer C is incorrect. Tension is measured to ensure the spring will close the valve correctly.

    Answer D is correct. It is not necessary to measure spring radius, as it will not change during use.

TASK H.6,
F.9

35. While monitoring the inlet air temperature sensor value with a scan tool, the technician finds that the value fluctuates when the throttle is opened and closed rapidly. This indicates:

    A.   The temperature sensor is operating normally.
    B.   The temperature sensor is shorted.
    C.   The temperature sensor is open.
    D.   The air filter is restricted.

    Answer A is correct. The sensor should change values when the throttle is opened suddenly.

    Answer B is incorrect. If the sensor was shorted, the value would not change, and a DTC would be set.

    Answer C is incorrect. If the sensor were open, the value would not change, and a DTC would be set.

    Answer D is incorrect. A restricted air filter cannot be located with this test method. A restricted air filter would cause low power and could be located using an air inlet restriction gauge.

36. An engine has a light ticking sound. Engine oil pressure is normal. Which of the following could be the cause?

    A. Worn rod bearings

    B. Worn main bearings

    C. Valves adjusted too loose

    D. Valves adjusted too tight

TASK A.4

Answer A is incorrect. Worn rod bearings usually cause a knocking noise accompanied by low oil pressure.

Answer B is incorrect. Worn main bearings usually cause a knocking noise accompanied by low oil pressure.

Answer C is correct. Loose valve adjustments can cause a light noise, and oil pressure will not be affected.

Answer D is incorrect. Tight valve adjustments can cause burnt valves and misfiring cylinders.

37. There is a light that resembles a gas cap illuminated on the dash. The technician should:

    A. Check the fuel level in the fuel tank

    B. Crain the fuel tank

    C. Inspect the gas cap

    D. Check fuel pump pressure

TASK F.2, D.12

Answer A is incorrect. Many vehicles have a low fuel level warning, usually in the form of text or a picture of a gas station pump, not a gas cap.

Answer B is incorrect. Some vehicles do have a water in fuel (WIF) indicator, which tells the operator to stop and drain the WIF separator; however, this light does not indicate fuel tank level.

Answer C is correct. The light indicates there is a leak in the EVAP system. The gas cap should be inspected; it is likely loose.

Answer D is incorrect. Some diesel engines do have a low fuel pressure warning, but this light is not illuminated based on fuel pump pressure.

38. While monitoring the mass airflow sensor value with a scan tool, the technician finds that the value fluctuates when the throttle is opened and closed rapidly. This indicates:

    A. The mass airflow sensor is operating normally.

    B. The mass airflow sensor is shorted.

    C. The mass airflow sensor is open.

    D. The sensor is dirty.

TASK H.6

Answer A is correct. The sensor should change values when the throttle is opened suddenly.

Answer B is incorrect. A shorted sensor would not register a change. A shorted sensor would set a DTC.

Answer C is incorrect. An open sensor would not register a change. An open sensor would set a DTC.

Answer D is incorrect. A dirty sensor will not register a change because the dirt prevents the sensor from recognizing the change in volume.

TASK E.1, F.9

39. A vehicle has poor fuel economy. Technician A says the cause could be a restricted air filter. Technician B says the cause could be a seized fan clutch. Who is correct?

    A. A only
    B. B only
    C. Both A and B
    D. Neither A nor B

    Answer A is incorrect. Technician B is also correct.

    Answer B is incorrect. Technician A is also correct.

    Answer C is correct. Both Technicians are correct. A restricted air filter can cause low power and poor fuel economy because the engine is starving for air. A seized fan clutch can make the fan run continuously, which causes the engine to work harder and consume extra fuel. Conversely, fan clutch failures can also result in the fan not spinning at required speeds, causing overheating but not affecting fuel economy.

    Answer D is incorrect. Both Technicians are correct.

TASK A.4

40. Engine balance shafts can be driven by all of the following EXCEPT:

    A. Gears
    B. Belts
    C. Chains
    D. Electric motors

    Answer A is incorrect. Engine balance shafts are often gear-driven.

    Answer B is incorrect. Engine balance shafts are often driven with a timing belt.

    Answer C is incorrect. Engine balance shafts are often driven with the timing chain.

    Answer D is correct. Engine balance shafts are never driven by an electric motor.

TASK G.3

41. An engine has a rough unstable idle and poor acceleration. Technician A says a stuck open exhaust gas recirculation (EGR) valve could be the cause. Technician B says a stuck closed PCV valve could be the cause. Who is correct?

    A. A only
    B. B only
    C. Both A and B
    D. Neither A nor B

    Answer A is correct. Only Technician A is correct. A stuck open EGR valve acts much like a vacuum leak, causing an unstable idle and poor acceleration.

    Answer B is incorrect. A stuck closed PCV valve will not affect idle quality. The valve is normally closed at idle. A stuck closed PCV valve could cause high crankcase pressure.

    Answer C is incorrect. Only Technician A is correct.

    Answer D is incorrect. Technician A is correct.

42. The specification for the cylinder bore on a given cast iron engine is 4.0125″. Listed below are the measurements for the engine:

TASK C.3

Cylinder #1	4.0126″	Cylinder #5	4.0126″
Cylinder #2	4.0125″	Cylinder #6	4.0126″
Cylinder #3	4.0126″	Cylinder #7	4.0125″
Cylinder #4	4.0127″	Cylinder #8	4.0127″

Which of the following would be the most likely repair procedure?

A. Bore the block
B. Sleeve the block
C. Reuse the block as is
D. Scrap the block

Answer A is incorrect. Typically, a block would be bored if the measurement were 0.002 greater than specification; this block would not be bored.

Answer B is incorrect. A block would be sleeved if it had deep scratches that could not be removed by light honing; this block would not be sleeved.

Answer C is correct. This block is well within typical wear limits and can be reused.

Answer D is incorrect. A block is only scrapped if there is wear or damage which is not economical to repair; this block does not need to be scrapped.

43. A truck will not start. When the fuel pump relay is jumped, the truck starts. Which of the following could be the cause?

TASK F.5

A. A failed fuel pump
B. A failed ignition control module
C. A failed fuel injector control module
D. A failed fuel pump relay

Answer A is incorrect. The engine would not start if the fuel pump has failed.

Answer B is incorrect. Jumping the fuel pump relay will not affect the ignition control module. The ignition control module is usually powered through a fuse.

Answer C is incorrect. Jumping the fuel pump relay would not affect the fuel injector control module. Jumping the power supply for the fuel injection control module would diagnose a faulty power supply for the fuel injection control module.

Answer D is correct. A failed fuel pump relay could be the cause. Jumping the relay would allow voltage to by-pass the failed relay and energize the fuel pump.

44. An engine has a single cylinder misfire; there is a popping sound from the exhaust when the misfire occurs. Which of the following could be the cause?

TASK A.5

A. A hole in a piston
B. A leaking intake gasket
C. A leaking thermostat housing gasket
D. A burnt exhaust valve

Answer A is incorrect. A hole in a piston could cause a misfire but would not likely cause a popping sound in the exhaust.

Answer B is incorrect. A leaking intake gasket could cause a misfire but would not likely cause a popping sound in the exhaust.

Answer C is incorrect. A leaking thermostat housing gasket would not cause a misfire.

Answer D is correct. A burnt exhaust valve could cause a misfire and a popping sound in the exhaust.

TASK D.2

45. An engine has suddenly lost oil pressure. Which of the following could be the cause?

A. The main bearings are worn.

B. The rod bearings are worn.

C. The oil pump pickup tube has come loose.

D. The piston rings have broken.

Answer A is incorrect. Worn main bearings can cause low oil pressure; however, it is typically a gradual process, not a sudden loss of oil pressure.

Answer B is incorrect. Worn rod bearings can cause low oil pressure; however, it is typically a gradual process, not a sudden loss of oil pressure.

Answer C is correct. If the oil pump pickup tube comes loose, the pump will suck air instead of oil, and the engine can experience a sudden loss of oil pressure.

Answer D is incorrect. Broken piston rings can cause oil consumption, not a sudden loss of oil pressure.

TASK H.6

46. All of the following can be used to test an accelerator pedal position (APP) sensor EXCEPT:

A. Oscilloscope

B. Ohmmeter

C. Ammeter

D. Scan tool

Answer A is incorrect. The voltage trace of an APP sensor can be tested with an oscilloscope to look for a momentarily open circuit.

Answer B is incorrect. An ohmmeter can be used to measure the resistance of the sensor.

Answer C is correct. There are no recognized tests using an ammeter on an APP sensor.

Answer D is incorrect. The scan tool can be an effective means of monitoring the APP sensor.

TASK H.3

47. While diagnosing a DTC, the technician finds a technical service bulletin that directly conflicts with service information for the vehicle. Which of the following should the technician do?

A. Follow the original service literature.

B. Follow the technical service bulletin.

C. Tell the customer to take the vehicle elsewhere.

D. Replace the part and see if that fixes the vehicle.

Answer A is incorrect. The technical service bulletin supersedes the service literature.

Answer B is correct. The technical service bulletin should have the most recent and up-to-date troubleshooting information.

Answer C is incorrect. The technician should follow the technical service bulletin.

Answer D is incorrect. The part should not be replaced until the diagnostic procedures determine it necessary.

48. An engine has low power. One cylinder has 40 percent leakage during a cylinder leakage test. All other cylinders have 15 percent leakage. Which of the following is true?

    A. This is a normal reading.

    B. This can be caused be retarded ignition timing.

    C. This can be caused by retarded camshaft timing.

    D. This can be caused by broken compression rings.

TASK A.9

    Answer A is incorrect. A reading of 40 percent is more than the normal specification of less than 20 percent and therefore indicates a problem with the cylinder.

    Answer B is incorrect. Retarded ignition timing can cause low power, but it will not cause excessive cylinder leakage.

    Answer C is incorrect. Retarded camshaft timing can cause low power, but it will not cause excessive cylinder leakage on one cylinder. It can affect cylinder leakage, but it would affect all cylinders that use that camshaft.

    Answer D is correct. Broken compression rings in one cylinder can cause low power and higher than normal leakage test results.

49. A vehicle has low power. Technician A says the cause could be retarded ignition timing. Technician B says the cause could be low ignition coil secondary voltage. Who is correct?

    A. A only

    B. B only

    C. Both A and B

    D. Neither A nor B

TASK E.1

    Answer A is incorrect. Technician B is also correct.

    Answer B is incorrect. Technician A is also correct.

    Answer C is correct. Both Technicians are correct. Retarded ignition timing can cause low power and poor fuel economy, because the spark occurs too late in the cycle. Low secondary coil output voltage can cause incomplete combustion, which causes the engine to have low power.

    Answer D is incorrect. Both Technicians are correct.

50. A truck dies while it is being driven; it will crank but not start. There are no DTCs. Which of the following should the technician check first?

    A. Throttle position sensor (TPS)

    B. APP

    C. EGR

    D. Fuel Level

TASK H.5

    Answer A is incorrect. It is likely that a failed TPS would set a DTC.

    Answer B is incorrect. A failed APP would likely set a DTC, and the vehicle may be reduced to an idle-only state.

    Answer C is incorrect. A failed EGR system could possibly cause the engine to stall at idle due to insufficient fresh air flow, but this would likely set a DTC.

    Answer D is correct. The technician should check the fuel level.

PREPARATION EXAM 2—ANSWER KEY

1.	A	**21.**	B	**41.**	C
2.	D	**22.**	C	**42.**	A
3.	C	**23.**	B	**43.**	A
4.	A	**24.**	C	**44.**	C
5.	A	**25.**	C	**45.**	B
6.	C	**26.**	D	**46.**	A
7.	D	**27.**	A	**47.**	A
8.	C	**28.**	C	**48.**	B
9.	B	**29.**	C	**49.**	D
10.	A	**30.**	B	**50.**	B
11.	C	**31.**	A		
12.	C	**32.**	C		
13.	A	**33.**	A		
14.	B	**34.**	B		
15.	C	**35.**	B		
16.	A	**36.**	C		
17.	D	**37.**	C		
18.	C	**38.**	B		
19.	C	**39.**	C		
20.	A	**40.**	A		

PREPARATION EXAM 2—EXPLANATIONS

TASK C.3

1. The specification for the cylinder bore on a given cast iron engine is 4.0225″. Listed below are the measurements for the engine:

Cylinder #1	4.0285″	Cylinder #5	4.0246″
Cylinder #2	4.0265″	Cylinder #6	4.0266″
Cylinder #3	4.0276″	Cylinder #7	4.0275″
Cylinder #4	4.0267″	Cylinder #8	4.0307″

Which of the following would be the most likely repair procedure?

A. Bore the block
B. Sleeve the block
C. Reuse the block as is
D. Scrap the block

Answer A is correct. Most likely this block would be bored 0.010 oversized and oversized pistons installed.

Answer B is incorrect. A block is sleeved when it has deep scratches that cannot be cleaned by boring.

Answer C is incorrect. The piston-to-cylinder wall clearance would be too great if used as is and would result in knocking and excessive oil consumption.

Answer D is incorrect. A block is scrapped when it is no longer cost effective to repair, as when the block is busted internally.

2. The first diagnostic step on most flow charts is:

   A. Verify the repair.
   B. Perform the repair.
   C. Identify the cause for concern.
   D. Verify concern.

TASK A.1

Answer A is incorrect. Verifying the repair would be done after the repair is complete.

Answer B is incorrect. Performing the repair would be done after the problem is identified.

Answer C is incorrect. Prior to identifying the cause for the concern, the concern must be verified.

Answer D is correct. The first step is usually to verify the concern.

3. Which of the following engine lubricating oils would typically be used in a gasoline-powered truck engine?

   A. 0W-20
   B. SAE40
   C. 10W-30
   D. 15W-40

TASK D.4

Answer A is incorrect. 0W-20 is usually only found in automotive gasoline engines. Even in that application, it is somewhat rare.

Answer B is incorrect. SAE40 is not currently recommended by any manufacturer for a gasoline-powered truck engine.

Answer C is correct. This is a typical oil recommendation for this application.

Answer D is incorrect. 15W-40 is typically used in diesel engines.

4. The normal resistance for the primary windings of an ignition coil is:

   A. 1 ohm
   B. 100 ohms
   C. 1,000 ohms
   D. 10,000 ohms

TASK E.5

Answer A is correct. The primary windings of an ignition coil typically have very low resistance. One ohm would be considered acceptable.

Answer B is incorrect. 100 ohms would be higher than normal for most ignition coils.

Answer C is incorrect. 1,000 ohms would be higher than normal for an ignition coil primary winding.

Answer D is incorrect. 10,000 ohms would be an extremely high primary winding resistance and would indicate a failed ignition coil.

TASK A.5,
B.6

5. An engine has a single cylinder misfire and there is a regularly occurring popping sound in the intake manifold. Which of the following could be the cause?

   A.   A broken exhaust pushrod
   B.   A broken intake pushrod
   C.   A faulty intake manifold pressure sensor
   D.   A faulty coolant temperature sensor

   Answer A is correct. A broken exhaust pushrod would prevent the exhaust from escaping during the exhaust stroke; thus, it would push back into the intake during the intake stroke, creating a regularly occurring popping sound in the intake manifold and a single cylinder misfire.

   Answer B is incorrect. A broken intake pushrod would cause a single cylinder miss, but it would not cause a popping sound in the intake manifold.

   Answer C is incorrect. A faulty intake manifold pressure sensor would not cause a single cylinder miss or a popping sound in the intake manifold. It could cause hard-starting or poor fuel mileage.

   Answer D is incorrect. A faulty coolant temperature sensor would not cause a single cylinder misfire or a popping sound in the intake manifold. It could cause a hard-start concern, poor fuel mileage or incorrect transmission shift points.

TASK A.5,
D.3

6. There is blue smoke coming from the exhaust of a gasoline-powered truck. Which of the following could be the cause?

   A.   A leaking radiator
   B.   A leaking oil cooler
   C.   Worn piston rings
   D.   Worn cam lobes

   Answer A is incorrect. A leaking radiator would most likely cause an external coolant leak.

   Answer B is incorrect. A leaking oil cooler would cause an external oil leak or an internal oil leak into the cooling system.

   Answer C is correct. Worn piston rings can allow oil into the combustion chamber. Burning oil produces blue exhaust smoke.

   Answer D is incorrect. Worn cam lobes can cause a misfire but would not likely cause blue exhaust smoke.

TASK F.6

7. A truck has low power. The fuel pressure is checked and is found to be lower than specification. Which of the following could be the cause?

   A.   A stuck closed pressure regulator
   B.   A stuck closed fuel injector
   C.   Low manifold vacuum
   D.   A restricted fuel filter

   Answer A is incorrect. A stuck closed regulator would not lower pressure; it would raise pressure.

   Answer B is incorrect. A stuck closed fuel injector would cause an engine misfire, but it would not have an effect on fuel pressure.

   Answer C is incorrect. Low manifold vacuum may raise fuel pressure if the regulator were vacuum controlled; however, it would not lower fuel pressure.

   Answer D is correct. A restricted fuel filter could lower fuel pressure and flow.

8. A vehicle runs acceptably while traveling at road speed but stalls when coming to a stop, is difficult to restart, and will not idle. Which of the following could be the cause?

TASK G.4

   A. Low fuel pressure
   B. A restricted fuel filter
   C. A stuck open exhaust gas recirculation (EGR) valve
   D. A stuck closed EGR valve

   Answer A is incorrect. Low fuel pressure would cause low power; this vehicle is said to operate acceptably at road speed.

   Answer B is incorrect. A restricted fuel filter would be more evident at road speed than at idle.

   Answer C is correct. A stuck open EGR valve would cause a loss of vacuum at idle and could cause this condition.

   Answer D is incorrect. A stuck closed EGR valve would not affect idle.

9. An engine is equipped with an AC voltage generator style crankshaft position sensor (CKP). This sensor can be bench checked with a:

TASK E.6

   A. Dwell meter
   B. Ohmmeter
   C. Oscilloscope
   D. Ammeter

   Answer A is incorrect. A dwell meter cannot be used to bench check this style of sensor. A dwell meter can be used to measure circuit on time.

   Answer B is correct. This sensor is bench checked by measuring the resistance. An ohmmeter can be used to measure the resistance and compare it to specification.

   Answer C is incorrect. An oscilloscope could be used to check the sensor on the vehicle, but not on the bench.

   Answer D is incorrect. An ammeter is not an effective tool to test this style of sensor. It is effective in measuring starter current draw.

10. There is black smoke coming from the exhaust of a gasoline-powered truck. Which of the following could be the cause?

TASK A.5

   A. A leaking fuel injector
   B. A leaking head gasket
   C. A leaking exhaust valve
   D. A leaking exhaust manifold

   Answer A is correct. A leaking fuel injector can cause over-fueling and black exhaust smoke.

   Answer B is incorrect. A leaking head gasket can allow oil or coolant into the combustion chamber. The resulting smoke is either blue or a steam like water vapor.

   Answer C is incorrect. A leaking exhaust valve will cause a puffing sound in the exhaust, but it will not likely cause black smoke.

   Answer D is incorrect. A leaking exhaust manifold will cause a noise and can allow exhaust gases to enter the cab, but it will not cause black smoke from the exhaust.

**TASK H.1,
A.11**

11. An engine has a surging problem at highway speeds. Engine operation is normal at idle and low speeds. Technician A says there may be a high resistance across the fuel pump relay. Technician B says that the inertia switch may have high resistance. Who is correct?

    A. A only

    B. B only

    C. Both A and B

    D. Neither A nor B

    Answer A is incorrect. Technician B is also correct.

    Answer B is incorrect. Technician A is also correct.

    Answer C is correct. Both Technicians are correct. A severe surging problem can be caused by high resistance in the relay or the inertia switch. High resistance in either item will cause low voltage at the fuel pump, which will result in fuel pump pressure and flow.

    Answer D is incorrect. Both Technicians are correct.

TASK A.5

12. There is the noticeable smell of exhaust in the cab and a backfire on engine deceleration. Which of the following could be the cause?

    A. A cracked intake manifold

    B. A leaking coolant temperature sensor

    C. A cracked exhaust manifold

    D. An internally leaking EGR valve

    Answer A is incorrect. A cracked intake manifold can cause a vacuum or coolant leak, not an exhaust leak.

    Answer B is incorrect. A leaking coolant temperature sensor can cause a visible coolant leak, not an exhaust leak.

    Answer C is correct. A cracked exhaust manifold can cause an exhaust leak and a backfire on deceleration.

    Answer D is incorrect. An internally leaking EGR valve can cause low power and a rough idle, but it would not cause the smell of exhaust inside the cab.

TASK F.4

13. A fuel pump has been replaced because of high current draw and low fuel pressure. In addition to replacing the fuel pump, the technician should also inspect:

    A. Voltage drop across the fuel pump electrical relay

    B. Applied voltage to the electronic control module (ECM)

    C. Voltage drop across the fuel injector electrical relay

    D. Applied voltage to the ignition control module

    Answer A is correct. A failing fuel pump can damage the relay contacts because of increased current flow. The relay should be tested to prevent a comeback.

    Answer B is incorrect. The fuel pump relay is not the power supply to the ECM. Therefore, increased current draw will have no effect on the voltage applied to the ECM

    Answer C is incorrect. The fuel pump relay does not normally supply voltage to the fuel injector control module. A fuel pump that draws too much current can damage the relay that supplies power to it. This circuit is not interconnected to the fuel injector power supply circuit.

    Answer D is incorrect. The fuel pump relay does not normally supply voltage to the ignition control module. The ignition control module does not interconnect with the fuel pump power supply circuit. Therefore, damage to one will not cause damage to the other.

14. Which tool would most likely be used to measure camshaft end-play on an in-block camshaft?

    A. Feeler gauge
    B. Dial indicator
    C. Inside micrometer
    D. Depth micrometer

    TASK C.5

    Answer A is incorrect. A feeler gauge can be used to measure end clearance in some applications; however, the most typical tool is a dial indicator.

    Answer B is correct. The dial indicator would be the most common tool to measure camshaft end-play.

    Answer C is incorrect. The inside micrometer is not able to measure end-play. It can be used to measure cylinder bore diameter.

    Answer D is incorrect. The depth micrometer is not usually used to measure end-play. It is used to measure valve spring installed height.

15. Which of the following would be the most normal intake manifold vacuum reading at idle?

    A. 5 in. Hg
    B. 10 in. Hg
    C. 17 in. Hg
    D. 27 in. Hg

    TASK A.6

    Answer A is incorrect. 5 in. Hg would be very low. This could be caused by an intake manifold leak.

    Answer B is incorrect. 10 in. Hg would be low. This could be caused by retarded ignition timing.

    Answer C is correct. 17 in. Hg would be a normal intake manifold vacuum reading at idle.

    Answer D is incorrect. 27 in. Hg would be much higher than normal. This reading would most likely indicate a failed vacuum gauge.

16. Technician A says spark plugs should be installed using a torque wrench. Technician B says spark plugs must be removed using a torque wrench. Who is correct?

    A. A only
    B. B only
    C. Both A and B
    D. Neither A nor B

    TASK E.4

    Answer A is correct. Only Technician A is correct. Spark plugs need to be installed to the correct torque to ensure proper seating and prevent damage to the spark plug and/or cylinder head.

    Answer B is incorrect. A torque wrench is not necessary to remove the spark plug.

    Answer C is incorrect. Only Technician A is correct.

    Answer D is incorrect. Technician A is correct.

TASK B.3

17. Which of the following would be used to measure cylinder head warpage?

   A. Micrometer
   B. Dial indicator
   C. Caliper and feeler gauge
   D. Straightedge and feeler gauge

   Answer A is incorrect. A micrometer can be used to measure crankshaft diameter, not cylinder head warpage.

   Answer B is incorrect. A dial indicator can be used to measure crankshaft end-play, not cylinder head warpage.

   Answer C is incorrect. A straightedge and a feeler gauge are the tools used to measure cylinder head warpage. A caliper can be used to measure outside, inside, or depth dimensions, not cylinder head warpage. A feeler gauge can also be used to measure clearances.

   Answer D is correct. A straightedge and a feeler gauge are the tools used to measure cylinder head warpage.

TASK G.6

18. A truck backfires on deceleration. Technician A says a faulty secondary air injection valve could be the problem. Technician B says an exhaust leak could be the problem. Who is correct?

   A. A only
   B. B only
   C. Both A and B
   D. Neither A nor B

   Answer A is incorrect. Technician B is also correct.

   Answer B is incorrect. Technician A is also correct.

   Answer C is correct. Both Technicians are correct. Either a faulty secondary air injection valve or an exhaust leak can allow oxygen into the exhaust system and the resulting backfire on deceleration.

   Answer D is incorrect. Both Technicians are correct.

TASK F.1

19. A truck is being diagnosed for a no-start concern. Technician A says that the level of fuel in the tank should be checked. Technician B says injector pulse width (PW) should be checked. Who is right?

   A. A only
   B. B only
   C. Both A and B
   D. Neither A nor B

   Answer A is wrong. Technician B is also correct.

   Answer B is wrong. Technician A is also correct.

   Answer C is correct. Both Technicians are correct. The fuel tank level should be checked on a no-start condition, and injector PW should also be checked.

   Answer D is wrong. Both Technicians are correct.

20. Which of the following can be used to help indicate the source of an engine oil leak?

    A. Smoke machine
    B. Pressure gauge
    C. Vacuum gauge
    D. Oscilloscope

TASK A.3

    Answer A is correct. A smoke machine can be used to apply smoke to the engine crankcase. The technician can watch for where the smoke leaks out to help identify the source of an engine oil leak.

    Answer B is incorrect. A pressure gauge can be useful in engine diagnosis to find low oil pressure; however, it would not help in locating the source of an engine oil leak.

    Answer C is incorrect. A vacuum gauge will show vacuum conditions within the engine, but it will not help identify the source of an engine oil leak.

    Answer D is incorrect. An oscilloscope would be used to locate problems within the ignition system. It would not be useful in locating an engine oil leak.

21. An engine has high hydrocarbon (HC) and carbon monoxide (CO) emissions. All of these defects could cause this problem EXCEPT:

    A. A leaking fuel injector
    B. An engine coolant temperature (ECT) sensor with low resistance
    C. An ECT sensor with high resistance
    D. A defective manifold absolute pressure (MAP) sensor

TASK H.1

    Answer A is incorrect. A leaking fuel injector can cause a rich air/fuel ratio and excessive HC and CO emissions.

    Answer B is correct. An ECT sensor with low resistance causes a lean air/fuel ratio, which causes a low HC and a high CO.

    Answer C is incorrect. An ECT sensor with high resistance will make the engine run rich and produce high HC and CO emissions.

    Answer D is incorrect. A MAP sensor can cause a rich air/fuel ratio and high HC and CO emissions.

22. Technician A says a plugged catalytic converter can be caused by a stuck open fuel injector. Technician B says a plugged catalytic converter can be caused by a misfiring cylinder. Who is correct?

    A. A only
    B. B only
    C. Both A and B
    D. Neither A nor B

TASK G.9,
F.13

    Answer A is incorrect. Technician B is also correct.

    Answer B is incorrect. Technician A is also correct.

    Answer C is correct. Both Technicians are correct. Excessive fuel in the exhaust system, whether caused by a stuck open fuel injector or a misfiring cylinder (both of which can dump raw fuel into the exhaust stream), can cause the converter substrate to overheat, melt and plug the converter.

    Answer D is incorrect. Both Technicians are correct.

TASK C.6

23. The crankshaft main bearing journal specification is 2.569″. The actual measurements are listed below:

Measurement #1 2.568″

Measurement #2 2.567″

Measurement #3 2.567″

Measurement #4 2.566″

Which of the following would be the most likely service procedure?

A. Reuse the crankshaft as is.

B. Turn the crankshaft 0.010″.

C. Turn the crankshaft 0.005″.

D. Weld the crankshaft and turn to original dimensions.

Answer A is incorrect. This crankshaft is worn beyond normal reuse guidelines.

Answer B is correct. This would be the most likely service procedure.

Answer C is incorrect. 0.005″ is not a normal service dimension.

Answer D is incorrect. This crankshaft could be welded and machined to original specification, but as that is a more costly procedure, it would not be the most likely.

TASK F.1, F.9

24. Which of these would be LEAST LIKELY to cause a no-start condition?

A. A dirty air filter

B. A restricted fuel filter

C. A faulty oxygen sensor

D. A faulty CKP

Answer A is incorrect. A dirty air filter can cause a no-start condition. It can cause the engine to receive so little air that it cannot ignite the air fuel mixture.

Answer B is incorrect. A restricted fuel filter can cause a no-start condition A severely restricted filter can result in insufficient fuel being delivered to the engine to start.

Answer C is correct. The oxygen sensor is used to adjust fuel trim after the engine is running. It is very unlikely that it would cause a no-start condition.

Answer D is incorrect. A faulty CKP can cause the ECM to fail to fire the spark plugs, resulting in a no-start condition.

TASK A.6

25. While using a vacuum gauge, the technician finds the intake manifold vacuum is low and steady on a gasoline engine. Which of the following is the most likely cause?

A. A leaking exhaust valve

B. A leaking intake valve

C. A restricted exhaust

D. A restricted air filter

Answer A is incorrect. A leaking exhaust valve can cause a pulsating reading.

Answer B is incorrect. A leaking intake valve can cause a pulsating reading.

Answer C is correct. A restricted exhaust can cause a low reading because if the exhaust cannot escape, then fresh air cannot be drawn in.

Answer D is incorrect. A restricted air filter will impede airflow and can cause a high intake manifold vacuum reading.

26. A vehicle is equipped with an ignition control module and a fuel injector control module. The ignition control module has shorted. Which of the following is the most likely cause of the failure?

TASK E.7

   A. A shorted fuel injector
   B. An open fuel injector
   C. An open coil pack
   D. A shorted coil pack

   Answer A is incorrect. A shorted fuel injector could possibly damage the fuel injector control module, but current for the fuel injector is not normally provided by the ignition module.

   Answer B is incorrect. An open fuel injector will result in no current flow and does not typically damage anything.

   Answer C is incorrect. An open circuit will not cause an increase in current flow. It will cause an engine misfire.

   Answer D is correct. A shorted coil pack can cause increased current draw and cause the ignition control module to fail.

27. The evaporative emissions control (EVAP) system is inoperative. Looking at the system monitors on the scan tool, the technician finds that the EVAP system monitor never runs. Technician A says a faulty engine temperature sensor signal could cause this. Technician B says a leaking vacuum hose at the EVAP solenoid could cause this. Who is correct?

TASK H.2

   A. A only
   B. B only
   C. Both A and B
   D. Neither A nor B

   Answer A is correct. Only Technician A is correct. The engine must reach the proper temperature for the powertrain control module (PCM) to operate the EVAP system. A faulty temperature sensor reading could prevent EVAP operation.

   Answer B is incorrect. A leaking vacuum hose may keep the EVAP system from functioning; however, it would not prevent the ECM from sending the command to the solenoid.

   Answer C is incorrect. Only Technician A is correct.

   Answer D is incorrect. Technician A is correct.

28. An intake manifold vacuum test is being performed on a gasoline-powered V8 truck engine. Which of the following would most likely be identified by a vacuum test?

TASK A.6

   A. A fouled spark plug on cylinder #6
   B. A leaking valve on cylinder #5
   C. A restricted exhaust
   D. A loose piston pin

   Answer A is incorrect. The vacuum test is not able to isolate a failure such as a fouled spark plug to an individual cylinder.

   Answer B is incorrect. The vacuum test is not able to isolate a failure such as a leaking valve to an individual cylinder.

   Answer C is correct. A restricted exhaust can be indicated by a lower than normal vacuum reading.

   Answer D is incorrect. A loose piston pin would cause a noise; however, it would not be indicated on a vacuum gauge.

TASK H.2

29. All of the following are true regarding the use of a scan tool EXCEPT:

   A. The engine should be warmed to determine if the PCM enters closed loop.

   B. The scan tool power adapter should be connected to a 12 V power supply.

   C. The PCM fuse should be removed prior to connecting to the scan tool.

   D. The year, model, and engine may be read automatically from the PCM.

   Answer A is incorrect. The engine should be warmed to normal operating temperature when using a scan tool to determine if the coolant temperature sensor is operating properly. The engine will also need to be warmed to determine if the PCM will enter closed loop.

   Answer B is incorrect. The power adaptor should be connected to the cigarette lighter.

   Answer C is correct. The PCM fuse does not need to be removed.

   Answer D is incorrect. The year, model, and engine information may be downloaded from the PCM to the scan tool automatically, or the technician may need to enter them manually.

TASK A.6

30. A technician is using a vacuum gauge to measure cranking vacuum. Which of the following would be the most normal reading?

   A. 1 in. Hg

   B. 4 in. Hg

   C. 10 in. Hg

   D. 18 in. Hg

   Answer A is incorrect. 1 in. Hg would be lower than normal and most likely indicate a leak.

   Answer B is correct. 4 in. Hg would be considered a normal cranking vacuum reading.

   Answer C is incorrect. 10 in. Hg would be higher than normal and most likely would be caused by a faulty gauge.

   Answer D is incorrect. 18 in. Hg would be a normal vacuum reading at idle.

TASK F.1

31. A truck requires extended cranking before it will start. Which of the following could be the cause?

   A. A faulty fuel pump relay

   B. Low battery voltage

   C. Carbon buildup on the pistons

   D. A restricted exhaust

   Answer A is correct. Some vehicles have a back-up fuel pump power supply circuit through the oil pressure switch. This circuit does not complete until the engine develops oil pressure.

   Answer B is incorrect. If a vehicle has low battery voltage, it will only get lower as the engine is cranked, so the truck will not eventually start after cranking.

   Answer C is incorrect. Carbon buildup on the pistons causes increased compression. It does not cause hard-starting.

   Answer D is incorrect. Restricted exhaust would become more evident, not less evident, with extended cranking, so the truck will not eventually start after cranking.

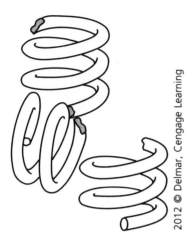

2012 © Delmar, Cengage Learning

32. Which of the following can cause the valve spring damage pictured above?

A. Excessive idling

B. A stuck open thermostat

C. Engine over-speeding

D. Incorrect piston-to-cylinder wall clearance

TASK B.4,
D.7

Answer A is incorrect. Idling will not cause a valve spring to break, although it can cause the valves to gum and stick.

Answer B is incorrect. A stuck open thermostat can cause an engine to operate too cool. However, it will not cause a valve spring to break.

Answer C is correct. Engine over-speeding can cause a valve spring to be over exerted and break.

Answer D is incorrect. Incorrect piston-to-cylinder wall clearance can cause a knock or piston seizure but will not cause a valve spring to break.

33. Technician A says a vacuum gauge can be used to help determine if the catalytic converter is clogged. Technician B says a fuel pressure gauge can be used to help determine if the catalytic converter is clogged. Who is correct?

A. A only

B. B only

C. Both A and B

D. Neither A nor B

TASK G.9

Answer A is correct. Only Technician A is correct. Intake manifold vacuum, which decreases as the engine is run, indicates the engine is not breathing properly and the exhaust is possibly restricted by a clogged catalytic converter.

Answer B is incorrect. A fuel pressure gauge may help find a faulty fuel pressure regulator or pump. However, it would not be helpful in determining if the converter is clogged.

Answer C is incorrect. Only Technician A is correct.

Answer D is incorrect. Technician A is correct.

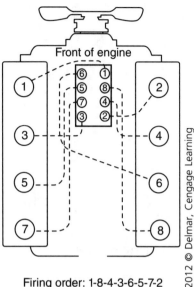

Front of engine

Firing order: 1-8-4-3-6-5-7-2

2012 © Delmar, Cengage Learning

TASK E.2

34. A vehicle has a firing order of 1-8-4-3-6-5-7-2, as seen in the cylinder arrangement pictured above. A diagnostic trouble code (DTC) has been set for a misfire on cylinders #5 and #7. All other cylinders are performing normally. Which of the following could be the cause?

   A.   Low fuel pump pressure
   B.   A failed head gasket
   C.   A worn exhaust cam lobe
   D.   A worn intake cam lobe

Answer A is incorrect. Low fuel pump pressure would not cause a misfire on only cylinders #5 and #7.

Answer B is correct. A failed head gasket between cylinders #5 and #7 could be the cause.

Answer C is incorrect. A worn exhaust cam lobe can cause a cylinder misfire, but it would not set a misfire code on two adjacent cylinders.

Answer D is incorrect. A worn intake cam lobe can cause erratic vacuum gauge readings, but it would not set a misfire code on two adjacent cylinders.

TASK A.6

35. The intake manifold vacuum is being measured during a road test. The vacuum gauge continues to drop the longer the vehicle is driven, and eventually the engine fails to accelerate and stalls. Which of the following could be the cause?

   A.   A restricted EGR passage
   B.   A restricted exhaust manifold
   C.   A leaking exhaust manifold
   D.   A leaking intake manifold

Answer A is incorrect. A restricted EGR passage could cause an engine ping due to increased cylinder temperatures, but it would not cause intake manifold vacuum to decrease.

Answer B is correct. A restricted exhaust can cause the vacuum to drop because the engine cannot breathe.

Answer C is incorrect. A leaking exhaust manifold can cause a noise, but it would not cause the symptoms indicated here.

Answer D is incorrect. A leaking intake manifold can cause a coolant leak or a low vacuum reading, but it would not cause the symptoms indicated here.

36. The engine is overheating. Which of the following is the most likely cause?

TASK D.6

    A.   Engine thermostat stuck open

    B.   Engine cooling fan stuck on

    C.   Engine thermostat stuck closed

    D.   Overtightened water pump belt

Answer A is incorrect. A stuck open thermostat will normally cause an engine to run colder than normal.

Answer B is incorrect. A constantly running cooling fan will cause excess engine noise and high fuel consumption, but not overheating.

Answer C is correct. A stuck closed engine thermostat can cause the coolant to fail to circulate and cause an engine overheating condition.

Answer D is incorrect. Overtightened water pump drive belts can cause premature pump failure but will not cause engine overheating.

37. Technician A says that a defective EGR valve may cause an engine to be hard to start. Technician B says that a defective ECT sensor may cause higher than normal emissions. Who is correct?

TASK H.3

    A.   A only

    B.   B only

    C.   Both A and B

    D.   Neither A nor B

Answer A is incorrect. Technician B is also correct.

Answer B is incorrect. Technician A is also correct.

Answer C is correct. Both Technicians are correct. A defective EGR valve can result in a vacuum leak and make the engine hard to start. Also, a faulty ECT can starve or flood the engine, which will cause incomplete combustion and an excess of hydrocarbons (HC) in the exhaust stream.

Answer D is incorrect. Both Technicians are correct.

38. An engine has spun a rod bearing. Besides repairing the crankshaft, what other service procedure must be performed?

TASK C.6

    A.   Bore the cylinders

    B.   Machine the connecting rod

    C.   Mill the head

    D.   Deck the block

Answer A is incorrect. The cylinders need to be bored only if they measure out of acceptable limits; the fact that the engine spun a bearing does not necessarily mean the cylinders will be out of specification.

Answer B is correct. If the bearing spun in the bore, the connecting rod will be damaged and must be machined.

Answer C is incorrect. The head should only be milled if damaged; the fact that the engine spun a bearing does not necessarily mean the head will be damaged.

Answer D is incorrect. The block should be decked only if damaged; the fact that the engine spun a bearing does not necessarily mean the block will be damaged.

Front of engine

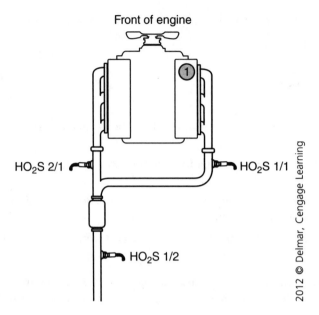

HO$_2$S 2/1

HO$_2$S 1/1

HO$_2$S 1/2

2012 © Delmar, Cengage Learning

TASK H.3

39. Refer to the above illustration. Which of the following can occur if the oxygen sensors HO$_2$S 1/1 and HO$_2$S 2/1 fail?

A. The GPS system will not function correctly.

B. The engine will run colder than normal.

C. The system will never go to closed loop.

D. The engine will not start.

Answer A is incorrect. The oxygen sensor will not affect the GPS system. The two are not interconnected.

Answer B is incorrect. A failed oxygen sensor will not significantly affect coolant temperature.

Answer C is correct. The PCM must recognize a good oxygen sensor signal to enter closed loop.

Answer D is incorrect. Oxygen sensor failure will not prevent the engine from starting.

TASK A.7

40. A cylinder power balance test is being performed on a gasoline-powered V8 truck engine. One cylinder reads lower than the others. Which of the following could be the cause?

A. A plugged fuel injector

B. A plugged exhaust system

C. A faulty oxygen sensor

D. A faulty oil pressure sensor

Answer A is correct. A plugged injector could reduce fuel delivery to one cylinder and cause a low power output on that cylinder.

Answer B is incorrect. A plugged exhaust system would affect the entire engine, not a single cylinder

Answer C is incorrect. A faulty oxygen sensor would affect the entire engine, not a single cylinder.

Answer D is incorrect. A faulty oil pressure sensor would affect the oil pressure gauge reading, not the power output of one cylinder.

41. A customer has had two failures of the EVAP system charcoal canister. Both canisters were flooded. All of the following could be the cause EXCEPT:

    A. Customer fueling habits
    B. A failed canister purge valve
    C. A plugged EGR passage
    D. A plugged vacuum line

TASK G.10

Answer A is incorrect. A customer who tops off the fuel tank excessively can cause the tank to overfill and the overage to fill the canister.

Answer B is incorrect. A failed purge valve could result in the purge cycle not completing; this would cause fuel vapors to continually collect in the canister, and eventually the canister would flood.

Answer C is correct. The EGR system is separate from the EVAP system and would not cause the canister to flood.

Answer D is incorrect. A plugged vacuum line, if connected to the EVAP system could prevent the system from purging and result in a flooded canister.

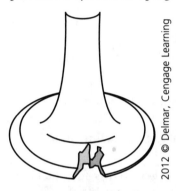

2012 © Delmar, Cengage Learning

42. All the exhaust valves in an engine are damaged, as pictured in the above illustration. Which of the following could be the cause?

    A. Adjusting the valves too tight
    B. Adjusting the valves too lose
    C. A faulty throttle position sensor (TPS)
    D. A leaking fuel injector

TASK B.4

Answer A is correct. Adjusting a valve too tight will result in less time for the valve to be in the seat, which could cause it to overheat and burn.

Answer B is incorrect. Adjusting the valve too loose can result in a noisy engine, but it will not cause the valve to overheat.

Answer C is incorrect. The TPS cannot cause burnt valves.

Answer D is incorrect. A leaking fuel injector can cause a cylinder misfire but could not cause a burnt valve.

43. Digital multi-meters (DMM) can be used on all the following scales EXCEPT:

    A. Milli-watts
    B. Amps
    C. Volts
    D. Ohms

TASK H.4

Answer A is correct. Milli-watts are not measured on a DMM.

Answer B is incorrect. Amps are measured on a DMM.

Answer C is incorrect. Volts are measured on a DMM.

Answer D is incorrect. Ohms are measured on a DMM.

TASK A.9

44. During a cylinder leakage test, shop air pressure should be set to:

A. 25 psi

B. 50 psi

C. 100 psi

D. 150 psi

Answer A is incorrect. 25 psi would not put enough pressure in the cylinder.

Answer B is incorrect. 50 psi would not be sufficient pressure to accurately tell if the rings and valves are sealing.

Answer C is correct. Although specifications may vary slightly, the most common specification would be 100 psi.

Answer D is incorrect. 150 psi would be too much pressure and would most likely damage the gauge.

TASK F.6

45. Which of the following would most likely set a multiple cylinder misfire DTC?

A. Low oil level voltage

B. Low fuel pressure

C. High fuel pressure

D. A faulty post-catalyst oxygen sensor

Answer A is incorrect. Low oil level can cause engine damage due to insufficient lubrication, but it will not set a misfire DTC.

Answer B is correct. Low fuel pressure could cause multiple cylinders to misfire, resulting in a multiple cylinder misfire DTC.

Answer C is incorrect. High fuel pressure can cause the engine to over-fuel and have increased emissions and poor fuel economy, but it will not cause a misfire DTC.

Answer D is incorrect. A faulty post-catalyst oxygen sensor may set a DTC for the catalyst, but it would not set a DTC for a misfire.

TASK H.4

46. Technician A says that an APP sensor can be checked with the DMM on the resistance scale. Technician B says that an APP sensor can be checked with the meter set on the AC voltage scale. Who is correct?

A. A only

B. B only

C. Both A and B

D. Neither A nor B

Answer A is correct. Only Technician A is correct. The sensor can be checked with the meter on the resistance scale.

Answer B is incorrect. The sensor generates a DC voltage. The AC voltage scale would not be used.

Answer C is incorrect. Only Technician A is correct.

Answer D is incorrect. Technician A is correct.

47. Engine coolant test strips can measure all of the following EXCEPT:
    A. Coolant temperature
    B. Coolant freeze protection
    C. Coolant pH
    D. Coolant acidity

TASK D.8

Answer A is correct. Test strips are not used to measure coolant temperature; this is done with a thermometer or pyrometer.

Answer B is incorrect. Coolant freeze protection can be measured with some test strips.

Answer C is incorrect. Coolant test strips can measure pH.

Answer D is incorrect. Coolant test strips can measure acidity.

48. Air is heard escaping the tailpipe during a cylinder leakage test. Which of the following could be the cause?
    A. A leaking intake valve
    B. A leaking exhaust valve
    C. A leaking head gasket
    D. A leaking exhaust manifold

TASK A.9

Answer A is incorrect. A leaking intake valve would allow air to escape from the throttle body.

Answer B is correct. A leaking exhaust valve would allow air to pass through the exhaust system and tailpipe during this test.

Answer C is incorrect. A leaking head gasket would usually cause bubbles in the radiator or air leaking into the adjacent cylinder during this test.

Answer D is incorrect. Exhaust manifold leaks are not located during the cylinder leakage test. Exhaust manifolds leaks are found with the engine running or with a smoke test.

49. A coil-on-plug (COP) ignition coil has low secondary voltage output. All of the following could be the cause EXCEPT:
    A. Low primary voltage
    B. High resistance in the primary windings
    C. A failed ignition coil
    D. A failed crankshaft sensor

TASK E.5

Answer A is incorrect. Low primary voltage can result in a weak magnetic field and low secondary voltage.

Answer B is incorrect. High resistance in the secondary windings can result in a lower than normal secondary voltage output.

Answer C is incorrect. A failed ignition coil can cause low secondary voltage output.

Answer D is correct. A failed crankshaft sensor may cause a no-start but would not cause low secondary voltage output.

50. What are valve spring shims used for?
    A. Correct weak valve springs
    B. Correct valve spring installed height
    C. Correct valve spring free length
    D. Correct valve spring squareness

TASK B.4

Answer A is incorrect. Weak springs must be replaced, not shimmed.

Answer B is correct. Shims are used to correct installed height after the seats and valves have been machined.

Answer C is incorrect. If free length is incorrect, the spring must be replaced.

Answer D is incorrect. If the valve spring is out of square, the spring must be replaced.

PREPARATION EXAM 3—ANSWER KEY

1. A	**21.** A	**41.** D
2. D	**22.** D	**42.** A
3. D	**23.** D	**43.** C
4. C	**24.** B	**44.** A
5. D	**25.** C	**45.** B
6. B	**26.** C	**46.** C
7. D	**27.** A	**47.** D
8. B	**28.** C	**48.** B
9. B	**29.** A	**49.** D
10. C	**30.** D	**50.** C
11. C	**31.** A	
12. A	**32.** C	
13. A	**33.** A	
14. C	**34.** B	
15. B	**35.** D	
16. D	**36.** B	
17. A	**37.** C	
18. C	**38.** A	
19. C	**39.** C	
20. D	**40.** C	

PREPARATION EXAM 3—EXPLANATIONS

TASK H.6

1. Technician A says that an exhaust gas recirculation (EGR) position sensor may be tested with either a voltmeter or an oscilloscope. Technician B says that if the powertrain control module (PCM) does not set an EGR position sensor fault code, there is no problem with the sensor. Who is correct?

   A. A only

   B. B only

   C. Both A and B

   D. Neither A nor B

   Answer A is correct. Only Technician A is correct. The EGR position sensor voltage output can be checked with a voltmeter or an oscilloscope.

   Answer B is incorrect. Even if the PCM has not set an EGR fault code, it is still possible that the system has had an in-range failure.

   Answer C is incorrect. Only Technician A is correct.

   Answer D is incorrect. Technician A is correct.

2. A customer has a low oil pressure concern. After verifying the complaint, what should the technician do next?

TASK D.1, D.12

    A. Replace the gauge in the dash.

    B. Replace the oil pressure sending unit.

    C. Measure the engine oil pressure with a master gauge.

    D. Check for proper engine oil level

Answer A is incorrect. The gauge should be replaced only if it is confirmed to be at fault.

Answer B is incorrect. The sending unit should only be replaced if it is confirmed to be at fault.

Answer C is incorrect. Measuring engine oil pressure with a master gauge is an important step in diagnosing a low oil pressure concern. However, it is not the best answer, because it is not the first test or check that should be done.

Answer D is correct. When an engine has a low oil pressure concern, the technician should make sure there is a correct amount of oil in the oil pan.

3. The normal resistance when measuring the primary to secondary windings of an ignition coil would be:

TASK E.5, A.11

    A. One ohm

    B. 100 ohms

    C. 1,000 ohms

    D. 10,000 ohms

Answer A is incorrect. One ohm would be considered shorted.

Answer B is incorrect. 100 ohms would be lower than normal for most ignition coils.

Answer C is incorrect. 1,000 ohms would be lower than normal for an ignition coil primary to secondary winding resistance.

Answer D is correct. The primary windings of an ignition coil typically have very high resistance. 10,000 ohms would be a normal primary to secondary winding resistance and would indicate an acceptable ignition coil.

4. A vehicle's electronic control module (ECM) must be replaced. Technician A says it is good practice to use a ground strap to ensure that there is no static discharge to the electronic circuit. Technician B says disconnecting the vehicle battery and touching bare metal on the vehicle prior to replacement will protect the vehicle from static discharge. Who is right?

TASK H.8

    A. A only

    B. B only

    C. Both A and B

    D. Neither A nor B

Answer A is incorrect. Technician B is also correct.

Answer B is incorrect. Technician A is also correct.

Answer C is correct. Both Technicians are correct. It is a good practice to protect electronic components from static discharge. A ground strap is an industry-accepted method to do this.

Answer D is incorrect. Both Technicians are correct.

TASK F.10

5. One intake manifold runner has a vacuum leak. Which of the following would most likely occur?

A. The engine would not start.

B. The engine would have excessive power.

C. The engine would not shut off.

D. The engine idle RPM would increase.

Answer A is incorrect. One intake manifold runner leaking would usually not prevent the engine from starting.

Answer B is incorrect. A leaking intake runner may cause low power, not excessive power.

Answer C is incorrect. A leaking intake runner will affect idle but will not prevent the engine from being shut off.

Answer D is correct. It is possible that engine idle would increase. The ECM will see a lean oxygen sensor due to the leak and add fuel to compensate; this can raise idle RPM.

TASK C.8, C.12

6. Plastigauge® is typically used to:

A. Measure piston-to-cylinder wall clearance

B. Measure bearing clearance

C. Measure camshaft end-play

D. Measure crankshaft end-play

Answer A is incorrect. Piston-to-cylinder wall clearance is calculated by measuring the bore and piston with micrometers

Answer B is correct. Bearing clearance is measured with Plastigauge.

Answer C is incorrect. Camshaft end-play is measured with a dial indicator.

Answer D is incorrect. Crankshaft end-play is measured with a dial indicator.

TASK B.7

7. The valves need to be adjusted on an overhead camshaft (OHC) engine that uses a shim style adjustment method. Technician A says the shim should be removed, then the clearance measured. Technician B says if the clearance is out of specification, the shim should be removed and ground to the appropriate dimension. Who is correct?

A. A only

B. B only

C. Both A and B

D. Neither A nor B

Answer A is incorrect. The clearance is measure with the shim installed. If the clearance is incorrect, the shim is removed and then replaced with a shim of appropriate thickness.

Answer B is incorrect. The shim must be installed when the clearance is measured.

Answer C is incorrect. Neither Technician is correct.

Answer D is correct. Neither Technician is correct. The clearance is measured with the shim in place. If the clearance is out of specification, then the shim is removed and replaced with one of the appropriate thickness.

8. A truck has a diagnostic trouble code (DTC) for a leaking evaporative emissions (EVAP) system. Which of the following tools would be used to locate the leak?

TASK G.11

   A. A nitrogen machine

   B. A smoke machine

   C. A cooling system pressure tester

   D. A compression gauge

Answer A is incorrect. A nitrogen machine is used to fill tires.

Answer B is correct. A smoke machine is used to help diagnosis leaks in low-pressure systems such as the EVAP system.

Answer C is incorrect. A cooling system pressure tester is used to locate leaks in the cooling system.

Answer D is incorrect. A compression tester is designed for a much higher pressure than found in the EVAP system.

9. A refractometer is used to measure coolant:

   A. Temperature

   B. Freeze protection

   C. pH

   D. Acidity

TASK D.8

Answer A is incorrect. Coolant temperature is measured with a thermometer.

Answer B is correct. Coolant freeze protection is measured with a refractometer.

Answer C is incorrect. Coolant pH is measured with test strips.

Answer D is incorrect. Coolant acidity is measured with test strips.

10. A truck is brought to the shop with a fuel-related DTC. Technician A says all codes should be recorded; then, they should be erased to see if any are active codes. Technician B says some scan tools list active and inactive for ease of interpretation. Who is correct?

TASK H.2

   A. A only

   B. B only

   C. Both A and B

   D. Neither A nor B

Answer A is incorrect. Technician B is also correct.

Answer B is incorrect. Technician A is also correct.

Answer C is correct. Both Technicians are correct. It is good practice to record DTCs, erase them from the PCM, and then drive the vehicle to see if any reoccur. Some scan tools do sort active and inactive codes.

Answer D is incorrect. Both Technicians are correct.

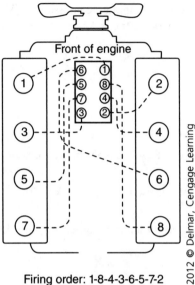

Front of engine

Firing order: 1-8-4-3-6-5-7-2

2012 © Delmar, Cengage Learning

TASK E.2

11. Refer to the above illustration. A vehicle has a firing order of 1-8-4-3-6-5-7-2. A DTC has been set for a misfire on cylinders #8 and #5. All other cylinders are performing normally. Which of the following could be the cause?

A. Low fuel pressure
B. A failed head gasket
C. A failed ignition coil
D. A worn intake cam lobe

Answer A is incorrect. Low fuel pressure would not cause a misfire on only cylinders #8 and #5.

Answer B is incorrect. Cylinders #8 and #5 are not adjacent to each other. When a head gasket fails between two cylinders next to each other, a misfire code can be set for both cylinders.

Answer C is correct. Cylinders #8 and #5 are companions; a failed ignition coil could affect both cylinders.

Answer D is incorrect. A worn intake cam lobe would not set a misfire code on two different cylinders. It can set a code for the cylinder that has the worn cam lobe.

TASK B.5, B.1

12. Cylinder head bolt torquing sequence usually begins:

A. In the middle
B. At the exhaust side
C. At the intake side
D. At the end

Answer A is correct. The torquing sequence usually starts in the middle to crush the gasket evenly and draw the head to the block squarely.

Answer B is incorrect. Starting at the exhaust side would not provide even gasket crush, which may result in gasket leakage.

Answer C is incorrect. Starting at the intake side would not provide even gasket crush, and a leaking head gasket could be the result.

Answer D is incorrect. Starting at the end would not provide even gasket crush. This may also cause the head to warp.

13. A truck has a DTC for a non-functioning catalytic converter. While comparing the oxygen sensors 2/1 and 2/2, the technician finds they are extremely similar in their voltage and response. Technician A says this indicates a non-functioning catalytic converter. Technician B says this indicates a non-functioning oxygen sensor 2/2. Who is correct?

TASK G.5

    A.　A only
    B.　B only
    C.　Both A and B
    D.　Neither A nor B

    Answer A is correct. Only Technician A is correct. The two oxygen sensors mentioned above should not be indicating the same values. If they are, it indicates that the catalyst is not working.

    Answer B is incorrect. A malfunctioning oxygen sensor 2/2 may send either high voltage or low voltage, but it would not send a voltage extremely similar to oxygen sensor 2/1. If both sensors are showing similar voltages, it is an indication that the catalytic converter has failed.

    Answer C is incorrect. Only Technician A is correct.

    Answer D is incorrect. Technician A is correct.

14. A technician is preparing to install a backpressure gauge to check the catalytic converter for restriction. Which of the following would be the correct location?

TASK G.9

    A.　In spark plug hole #6
    B.　In spark plug hole #8
    C.　In oxygen sensor hole 1/1
    D.　In oxygen sensor hole 2/1

    Answer A is incorrect. Installing the gauge in spark plug hole #6 would result in taking a compression reading on cylinder #6.

    Answer B is incorrect. Installing the gauge in spark plug hole #8 would result in measuring the compression on cylinder #8.

    Answer C is correct. The correct location is in the oxygen sensor hole prior to the converter: hole 1/1.

    Answer D is incorrect. Oxygen sensor hole 2/1 is after the converter; this would be after any pressure buildup caused by a restricted converter and would not be helpful information.

15. All of the following could cause a single cylinder misfire EXCEPT:

TASK F.7

    A.　A fouled spark plug
    B.　A dirty throttle plate
    C.　A stuck open fuel injector
    D.　A stuck closed fuel injector

    Answer A is incorrect. A fouled spark plug can cause one cylinder to misfire.

    Answer B is correct. A dirty throttle plate can cause rough idle and stalling, but it would not cause a single cylinder misfire.

    Answer C is incorrect. A stuck open fuel injector can flood the cylinder and cause a single cylinder misfire.

    Answer D is incorrect. A stuck closed fuel injector can starve the cylinder and cause a single cylinder misfire.

TASK A.9

16. During a cylinder leakage test, bubbles are found in the radiator. Which of the following is indicated?

   A.   A restricted radiator
   B.   A restricted head gasket
   C.   A leaking radiator
   D.   A leaking head gasket

   Answer A is incorrect. A restricted radiator is not located during a cylinder leakage test.

   Answer B is incorrect. A restricted head gasket is not located during a cylinder leakage test.

   Answer C is incorrect. A leaking radiator can be located during a cooling system pressure test, but it would not cause bubbles.

   Answer D is correct. A leaking head gasket can allow air to be introduced into the cooling system during a cylinder leakage test, resulting in bubbles in the radiator.

TASK C.9

17. The color of Plastigauge determines its:

   A.   Diameter
   B.   Length
   C.   Temperature
   D.   Age

   Answer A is correct. The color of Plastigauge indicates its diameter and, therefore, which bearing clearance specification set it should be used with.

   Answer B is incorrect. Plastigauge comes in long strips and is cut to length by the technician.

   Answer C is incorrect. Temperature has no bearing on Plastigauge.

   Answer D is incorrect. As Plastigauge ages, it becomes less accurate. Therefore, the technician should use new Plastigauge.

TASK G.5

18. A DTC has been set for the EGR system. Technician A says plugged passage tubes connecting the EGR pressure sensor could be the cause. Technician B says a failed EGR pressure sensor could be the cause. Who is correct?

   A.   A only
   B.   B only
   C.   Both A and B
   D.   Neither A nor B

   Answer A is incorrect. Technician B is also correct.

   Answer B is incorrect. Technician A is also correct.

   Answer C is correct. Both Technicians are correct. A DTC for the EGR system can be caused by clogged tubes connecting to the sensor or by a failed sensor. Either failure would result in the ECM not reading the correct pressure change when the EGR valve is opened.

   Answer D is incorrect. Both Technicians are correct.

TASK A.8

19. Which of the following is a normal compression reading during a compression test?

   A.   50 psi
   B.   75 psi
   C.   125 psi
   D.   225 psi

   Answer A is incorrect. 50 psi would be too low to support combustion.

   Answer B is incorrect. 75 psi may support combustion but would result in low power contribution from that cylinder.

   Answer C is correct. Normal compression readings will vary from engine to engine but usually fall between 100 to175 psi.

   Answer D is incorrect. 225 psi would be excessively high and would likely result in detonation.

20. A coil on plug ignition coil has failed and must be replaced. Technician A says the ignition coil on the companion cylinder must also be replaced. Technician B says the fuel injector must also be replaced. Who is correct?

    A. A only
    B. B only
    C. Both A and B
    D. Neither A nor B

TASK E.5

    Answer A is incorrect. There is no reason to replace the coil on the companion cylinder.

    Answer B is incorrect. There is no reason to replace the accompanying injector.

    Answer C is incorrect. Neither Technician is correct.

    Answer D is correct. Neither Technician is correct. In a coil-on-plug ignition system, there is one coil per cylinder; therefore, when a coil-on-plug ignition system has a coil failure, it only affects that single cylinder.

21. Which of the following would be performed during a power balance test?

    A. Remove the spark from a cylinder.
    B. Remove the coolant from a cylinder.
    C. Remove the intake air from a cylinder.
    D. Remove the exhaust from a cylinder.

TASK A.7

    Answer A is correct. Usually spark or fuel is removed from a cylinder during a power balance test.

    Answer B is incorrect. There should not be any coolant in the cylinder.

    Answer C is incorrect. There is no practical way to remove the intake air from the cylinder.

    Answer D is incorrect. There is no practical way to remove the exhaust from the cylinder.

22. All cylinders read lower than normal on a cranking compression test. Which of the following could be the cause?

    A. Excessive cranking speed
    B. A leaking intake valve
    C. A leaking exhaust valve
    D. An out-of-time camshaft

TASK A.8

    Answer A is incorrect. Excessive cranking speed would not cause low compression. However, excessive cranking speed can indicate low compression.

    Answer B is incorrect. A leaking intake valve would not affect all cylinders. It would only affect the cylinder where the valve is located.

    Answer C is incorrect. A leaking exhaust valve would not affect all cylinders. It would only affect the cylinder where the valve is located.

    Answer D is correct. An out-of-time camshaft would affect all cylinders and could cause all cylinders to read low during this test.

23. An exhaust manifold is leaking at the cylinder head. Technician A says the manifold must be replaced. Technician B says the cylinder head must be replaced. Who is correct?

    A. A only

    B. B only

    C. Both A and B

    D. Neither A nor B

    Answer A is incorrect. An exhaust leak can be caused by a warped exhaust manifold; however, if the manifold is warped, it may be possible to resurface it.

    Answer B is incorrect. A cylinder head that is warped at the exhaust manifold mounting surface can cause an exhaust leak. However, if the cylinder gasket surface is warped, it may be possible to resurface it.

    Answer C is incorrect. Neither Technician is correct.

    Answer D is correct. Neither Technician is correct. When there is an exhaust leak at the exhaust manifold sealing surface, it may be possible to resurface the related components and not have to replace them.

24. Piston ring end gap is measured with:

    A. With the rings on the piston

    B. With the rings in the cylinder bore

    C. With a dial indicator

    D. After the engine is assembled

    Answer A is incorrect. Piston ring end gap is measured with the ring in the bore. The ring must be off the piston so there is room to install the feeler gauge.

    Answer B is correct. Piston ring end gap is measured with the rings in the bore. The rings can be squared in the bore with the head of the piston. The rings must be in the bore to measure end gap.

    Answer C is incorrect. A feeler gauge is used to measure end gap. A dial indicator is used to measure crankshaft end-play.

    Answer D is incorrect. The measurement must be taken prior to assembling the engine.

25. A customer is concerned that coolant must be added to the cooling system. The technician finds no evidence of an external leak. A cylinder leakage test reveals bubbles in the coolant. Which condition is indicated?

    A. Leaking radiator

    B. Leaking oil cooler

    C. Leaking head gasket

    D. Leaking throttle plate gasket

    Answer A is incorrect. A leaking radiator will cause external leaks and is best confirmed using a cooling system pressure test.

    Answer B is incorrect. A leaking oil cooler will cause the coolant and oil to mix, but will not cause bubbles in the coolant during a cylinder leakage test.

    Answer C is correct. Bubbles in the coolant during a cylinder leakage test can be caused by a leaking head gasket.

    Answer D is incorrect. A leaking throttle plate gasket can cause a coolant leak if the throttle plate is warmed by coolant, but this will not cause bubbles in the radiator during a cylinder leakage test.

26. Where would a technician locate the most recent software upgrades for an engine ECM?

    A. Service manuals

    B. CD-ROMs

    C. The original equipment manufacturer (OEM) service website

    D. Trade publications

**TASK H.9,
A.13**

Answer A is incorrect. Service manuals are great resources for repair procedures, but not software upgrades.

Answer B is incorrect. Most service literature is provided on CD-ROMs, but not OEM software upgrades.

Answer C is correct. The most recent software upgrades for an engine ECM are available by accessing the OEM service website.

Answer D is incorrect. Trade publications are not used for ECM software upgrades.

27. The air intake system is being inspected for leaks. Technician A says a smoke machine may be used. Technician B says starting fluid may be used. Who is correct?

    A. A only

    B. B only

    C. Both A and B

    D. Neither A nor B

TASK F.10

Answer A is correct. Only Technician A is correct. A smoke machine can be used to fill the intake system with smoke and watch for the smoke trails, which would indicate a leak.

Answer B is incorrect. It is not a recommended practice, due to safety reasons, to spray starting fluid around the intake system to try to locate a leak.

Answer C is incorrect. Only Technician A is correct.

Answer D is incorrect. Technician A is correct.

28. A vehicle has the check engine light on. The cruise control, automatic door locks, and speedometer are not functioning. Which of the following is the most likely cause?

    A. A faulty PCM

    B. A faulty oxygen sensor

    C. A faulty vehicle speed sensor

    D. A faulty wheel speed sensor

TASK H.10

Answer A is incorrect. A faulty PCM can cause the check engine light to come on. However, with this set of symptoms, it is not the most likely cause.

Answer B is incorrect. A faulty oxygen sensor would not cause the door locks or speedometer concerns.

Answer C is correct. All of these systems need input from the vehicle speed sensor.

Answer D is incorrect. Wheel speed sensors typically cause ABS brake system failure codes.

TASK B.8,
C.10

29. An engine with an OHC has had the deck milled. Technician A says this can affect camshaft timing. Technician B says this can affect fuel pressure. Who is correct?

A. A only

B. B only

C. Both A and B

D. Neither A nor B

Answer A is correct. Only Technician A is correct. When the block deck is milled, the center lines of the crankshaft and camshaft will be drawn closer together, which will affect the positioning of the timing chain and can affect camshaft timing.

Answer B is incorrect. Though it will raise cylinder pressure, milling the deck will have no affect on fuel pressure.

Answer C is incorrect. Only Technician A is correct.

Answer D is incorrect. Technician A is correct.

TASK A.4,
C.14

30. An engine has a knock that is not load dependent and can only be heard at 1000 rpm, 1700 rpm, 2400 rpm, and 3100 rpm. Which of the following is the most likely cause?

A. Rod bearing

B. Main bearing

C. Piston pin

D. Vibration damper

Answer A is incorrect. A rod bearing may not be evident at all RPMs. However, it is load dependent.

Answer B is incorrect. A main bearing knock will not appear at only certain RPMs. It will be evident throughout the RPM range.

Answer C is incorrect. A piston pin may not be evident at all RPMs. However, it is load dependent.

Answer D is correct. A knock that is evident at only certain RPMs and not load dependent is most likely caused by the vibration damper.

TASK E.6

31. The acronym CMP is used for:

A. Camshaft Position Sensor

B. Crankcase Pressure Sensor

C. Crankshaft Position Sensor

D. Cylinder Pressure Sensor

Answer A is correct. The camshaft position sensor's acronym is CMP.

Answer B is incorrect. CMP is the acronym for Camshaft Position Sensor. There is no universal acronym for crankcase pressure sensor.

Answer C is incorrect. The crankshaft position sensor typically uses the acronym CKP.

Answer D is incorrect. CMP is the acronym for Camshaft Position Sensor. There is no universal acronym for a cylinder pressure sensor.

32. An engine has an oil leak. Technician A says fluorescent dye can be used to locate the source of the leak. Technician B says smoke can be used to locate the source of the leak. Who is correct?

TASK A.3, C.1

A. A only

B. B only

C. Both A and B

D. Neither A nor B

Answer A is incorrect. Technician B is also correct.

Answer B is incorrect. Technician A is also correct.

Answer C is correct. Using Both Technicians are correct. Fluorescent dye is a good method for locating engine oil leaks. A smoke generator can also be used to fill the crankcase with smoke and see where the smoke exits.

Answer D is incorrect. Both Technicians are correct.

33. Which of the following would be a common metric bolt designation?

TASK B.2

A. M8 × 1.0 mm

B. M8 × 18 tpi

C. 3/8 × 18 tpi

D. 3/8 × 1.5 mm

Answer A is correct. This is a common metric bolt designation. This indicates the bolt is 8mm in diameter with 1.0 mm distance between the thread.

Answer B is incorrect. This answer mixes metric and US English designations.

Answer C is incorrect. This is a US SAE standard bolt designation. This bolt is 3/8 inch in diameter and has 18 threads per inch.

Answer D is incorrect. This answer mixes metric and US English bolt designation.

34. An engine which has high oil consumption also has high crankcase pressure. Which of the following is the most likely cause?

TASK A.5

A. Worn intake valve guides

B. Worn piston rings

C. Worn exhaust valve guides

D. Worn camshaft lobes

Answer A is incorrect. Worn intake valve guides can cause oil consumption; however, it would not cause high crankcase pressure.

Answer B is correct. Worn piston rings will cause high oil consumption and high crankcase pressure.

Answer C is incorrect. Worn exhaust valve guides may cause increased oil consumption but usually will not cause higher than normal crankcase pressure.

Answer D is incorrect. Worn camshaft lobes will cause low power and poor intake manifold readings. They will cause neither high oil consumption nor high crankcase pressure.

TASK F.6

35. A vacuum-controlled fuel pressure regulator is being tested. What is the correct procedure?

   A. With the engine shut off, monitor fuel pressure while removing the vacuum hose.

   B. Remove the regulator and submerge it in fuel.

   C. Remove the regulator from the engine and apply vacuum.

   D. With the engine running, monitor fuel pressure while removing the vacuum hose.

   Answer A is incorrect. Since there is no engine vacuum with the engine off, removing the hose would have no effect.

   Answer B is incorrect. There is no reason to remove the regulator and submerge it in fuel.

   Answer C is incorrect. If the regulator is removed from the engine, it will not be possible to determine if the fuel pressure has changed.

   Answer D is correct. The correct test method is to monitor fuel pressure with the engine running. Remove the vacuum hose and fuel pressure should increase.

TASK G.3

36. There is a DTC set for an inoperative EGR valve. When the technician opens the EGR valve manually with the engine running, the engine stumbles and dies. Which of the following could be the cause of the inoperative EGR valve code?

   A. Restricted EGR passages

   B. Failed EGR pressure sensor

   C. Clogged EVAP canister converter

   D. Punctured catalytic converter

   Answer A is incorrect. If the EGR passages were restricted, the engine would not have stumbled when the technician manually opened the valve.

   Answer B is correct. A failed EGR sensor would not send the correct signal to the ECM, and an EGR DTC would be set.

   Answer C is incorrect. A clogged EVAP canister would not cause an EGR DTC. It could cause an EVAP system DTC.

   Answer D is incorrect. A punctured catalytic converter would set a catalytic converter DTC.

TASK H.6

37. Technician A says a defective ECT sensor may cause the engine to be hard to start. Technician B says a defective ECT sensor may cause the engine to fail an emissions test. Who is correct?

   A. A only

   B. B only

   C. Both A and B

   D. Neither A nor B

   Answer A is incorrect. Technician B is also correct.

   Answer B is incorrect. Technician A is also correct.

   Answer C is correct. Both Technicians are correct. A faulty ECT can cause the ECM to send the incorrect amount of fuel, which could result in an engine that is either over-fueling or under-fueling, making it hard to start. This over-fueling can also cause the engine to have higher than normal emission levels.

   Answer D is incorrect. Both Technicians are correct.

38. An engine has normal intake manifold vacuum at idle. When the vehicle is driven, the intake manifold vacuum continually gets lower, and eventually the engine stalls. Which test should the technician perform next?

TASK A.6, F.13

   A. Exhaust restriction
   B. Cranking compression
   C. Running compression
   D. Air filter restriction

   Answer A is correct. Restricted exhaust can cause an engine to lose its ability to breathe. The exhaust does not exit, so fresh air cannot enter; therefore, the intake manifold vacuum will drop, and the engine can eventually stall.

   Answer B is incorrect. This engine has good vacuum at idle; it is very unlikely that cranking compression will indicate low compression. Additionally, low compression would not cause the vacuum to fall off during a test drive.

   Answer C is incorrect. A running compression test is very useful for locating valve and cam problems. However, since vacuum is good at idle but falls slowly while driving, camshaft and valve problems are not indicated. Camshaft and valve problems are usually identified by regular downward pulses on a vacuum gauge.

   Answer D is incorrect. Air filter restriction will cause higher than normal vacuum readings, because the engine cannot get air.

39. Technician A says fuel injectors can be cleaned on the engine. Technician B says fuel injectors can be cleaned on a test bench. Who is correct?

TASK F.8

   A. A only
   B. B only
   C. Both A and B
   D. Neither A nor B

   Answer A is incorrect. Technician B is also correct.

   Answer B is incorrect. Technician A is also correct.

   Answer C is correct. Both Technicians are correct. Methods have been developed to clean the injectors both on the engine and on a test bench.

   Answer D is incorrect. Both Technicians are correct.

40. During a power balance test, two adjacent cylinders are found to have low power. Which of the following could be the cause?

TASK A.7

   A. A restricted fuel filter
   B. A restricted air filter
   C. Switched spark plug wires
   D. Switched oxygen sensor connectors

   Answer A is incorrect. A restricted fuel filter would not affect only two cylinders.

   Answer B is incorrect. A restricted air filter would not affect only two cylinders.

   Answer C is correct. Switched spark plug wires could easily cause low power output on two adjacent cylinders.

   Answer D is incorrect. It would be difficult to switch oxygen sensor connectors; however, if it did occur, it would not affect only two cylinders.

TASK A.7

41. A power balance test is being performed on a gasoline-powered V10 truck engine. The front six cylinders have good power output. The rear four cylinders have less than adequate power output. Which of the following could be the cause?

A.   A restricted air filter

B.   Low engine oil level

C.   Low transmission fluid level

D.   Restricted coolant passages

Answer A is incorrect. A restricted air filter would not affect only the four rear cylinders.

Answer B is incorrect. An engine should not be operated with a low oil level; however, low engine oil levels typically do not cause low output during a cylinder power balance test.

Answer C is incorrect. Transmission fluid level is not a concern during a power balance test.

Answer D is correct. Restricted coolant passages could cause the rear of an engine to overheat. This can cause the compression rings on those pistons to lose tension, resulting in low power output.

TASK H.1

42. Technician A says that closed-loop operation occurs when the engine is at operating temperature, and the PCM uses the information from the oxygen sensor to manage the air/fuel ratio. Technician B says that while in closed-loop operation, injector pulse width (PW) is measured in seconds. Who is correct?

A.   A only

B.   B only

C.   Both A and B

D.   Neither A nor B

Answer A is correct. Only Technician A is correct. In closed-loop operation, the engine is at operating temperature, and the PCM uses the information from the oxygen sensor to control the air/fuel ratio.

Answer B is incorrect. Injector PW is measured in milliseconds.

Answer C is incorrect. Only Technician A is correct.

Answer D is incorrect. Technician A is correct.

TASK A.8

43. An engine has low power. A cranking compression test reveals two adjacent cylinders have low compression. Which additional test will best help isolate the problem?

A.   Intake manifold vacuum

B.   Exhaust restriction

C.   Cylinder leakage

D.   Running compression

Answer A is incorrect. An intake manifold test works well to find intake restricted exhaust or late valve timing. It would not be useful to isolate a blown head gasket to a cylinder location.

Answer B is incorrect. An exhaust restriction test will not help confirm a leaking head gasket between two adjacent cylinders. However, it is effective in finding a restricted catalytic converter.

Answer C is correct. A cylinder leakage test will identify a leaking head gasket between two cylinders. When air is introduced into the cylinder, it will exit the cylinder next to it.

Answer D is incorrect. A running compression test is useful for finding worn cam lobes or broken valve springs, but it will not definitely confirm a leaking head gasket between two cylinders.

44. A cylinder leakage test is being performed on a truck. Which of the following would indicate an acceptable amount of cylinder leakage?

TASK A.9

   A. 15 percent

   B. 30 percent

   C. 45 percent

   D. 60 percent

Answer A is correct. Usually less than 20 percent is considered an acceptable reading.

Answer B is incorrect. A 30 percent reading can be caused by a leaking intake valve.

Answer C is incorrect. A 45 percent reading would result in low power from the engine.

Answer D is incorrect. A 60 percent reading would likely result in a cylinder misfire.

45. Technician A says a surface comparator is used to determine the surface hardness of metal. Technician B says a surface comparator can be used to check the machined surface of a cylinder block. Who is correct?

TASK C.2, B.1

   A. A only

   B. B only

   C. Both A and B

   D. Neither A nor B

Answer A is incorrect. A Rockwell gauge is used to test metal hardness.

Answer B is correct. Only Technician B is correct. A comparator is used to check the finish of a machine surface. This tool would be used to insure that the cylinder block and/or cylinder head has the appropriate finish for the style of head gasket being installed.

Answer C is incorrect. Only Technician B is correct.

Answer D is incorrect. Technician B is correct.

46. There is no spark at any of the spark plugs on a truck equipped with a coil-on-plug ignition system. Which of the following could be the cause?

TASK E.1

   A. A faulty mass airflow sensor

   B. A faulty ignition coil

   C. A failed CKP

   D. A failed EGR pressure sensor

Answer A is incorrect. The mass airflow sensor can cause rough idle and poor acceleration, but it would not prevent spark at the spark plugs.

Answer B is incorrect. This vehicle has an ignition coil for each cylinder. A single failed coil would not affect the other cylinders.

Answer C is correct. A failed CKP would not be able to relate the crankshaft position to the ECM; the result would be no spark at any spark plug.

Answer D is incorrect. A failed EGR pressure sensor may cause the EGR system to operate incorrectly. However, that would not prevent a spark at the spark plugs.

47. A truck has a factory fill of orange DEX-COOL® as the coolant. Technician A says the coolant can be topped off with green ethylene glycol. Technician B says the coolant can be topped off with European Pink. Who is correct?

    A. A only
    B. B only
    C. Both A and B
    D. Neither A nor B

    Answer A is incorrect. The coolant should be topped off with only DEX-COOL; green ethylene glycol should not be mixed in.

    Answer B is incorrect. The coolant should only be topped off with DEX-COOL; European Pink should not be mixed in.

    Answer C is incorrect. Neither Technician is correct.

    Answer D is correct. Neither Technician is correct. Mixing coolant types can result in thick coolant, which plugs coolant passages and results in engine overheating.

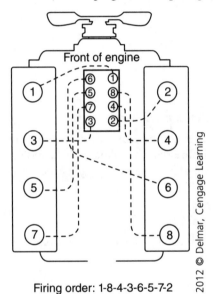

Firing order: 1-8-4-3-6-5-7-2

48. Refer to the above illustration. A cylinder power balance test has been performed on an engine. The firing order of the engine is 1-8-4-3-6-5-7-2. Cylinders #1 and #8 show lower RPM drop than the others. Which of the following is the most likely cause?

    A. Restricted cooling system
    B. Incorrect spark plug wire routing
    C. Leaking intake valves
    D. Leaking compression rings

    Answer A is incorrect. A restricted cooling system could cause the engine's rear cylinders to get insufficient coolant flow, result in weak performance. However, on this engine one weak cylinder is located in the front, and the other is located in the rear.

    Answer B is correct. Since the two affected cylinders are next to each other in the distributor cap, it is very likely that the spark plug wires have been switched.

    Answer C is incorrect. Leaking intake valves can cause weak performance during a cylinder power balance test; however, it is not the most likely cause, since the two cylinders affected are located next to each other in the distributor cap.

    Answer D is incorrect. Leaking compression rings can cause poor readings on a cylinder power balance test. However, since the two cylinders affected are located next to each other in the distributor cap, it is not the most likely cause.

49. Spark plugs should be installed with a:

A. 3/8″ dr. air impact wrench

B. 3/8″ dr. butterfly

C. 3/8″ dr. electric impact

D. Torque wrench

TASK E.4

Answer A is incorrect. Using an air impact wrench can result in stripping the threads.

Answer B is incorrect. A butterfly is useful to quickly tighten fasteners. However, it is not appropriate to use where installation torque is critical to proper sealing.

Answer C is incorrect. An electric impact is a popular tool because it is cordless, but it will not properly torque the spark plug.

Answer D is correct. The only correct method to install spark plugs is with a torque wrench.

50. OBD-II provides for standardization of all of the following EXCEPT:

A. Communication protocols

B. Diagnostic connectors

C. Fuel calibrations

D. DTCs

TASK H.2

Answer A is incorrect. OBD-II sets communication protocols.

Answer B is incorrect. OBD-II establishes connector standards.

Answer C is correct. Fuel calibrations vary widely according to vehicle configurations and operating conditions.

Answer D is incorrect. OBD-II provides standardized DTCs.

PREPARATION EXAM 4—ANSWER KEY

1. C		**21.** A		**41.** C	
2. A		**22.** A		**42.** A	
3. C		**23.** D		**43.** A	
4. D		**24.** B		**44.** C	
5. C		**25.** B		**45.** B	
6. D		**26.** C		**46.** C	
7. B		**27.** A		**47.** B	
8. B		**28.** D		**48.** C	
9. C		**29.** C		**49.** C	
10. A		**30.** C		**50.** D	
11. C		**31.** B			
12. B		**32.** A			
13. D		**33.** C			
14. C		**34.** A			
15. B		**35.** B			
16. A		**36.** B			
17. C		**37.** A			
18. D		**38.** C			
19. A		**39.** C			
20. D		**40.** B			

PREPARATION EXAM 4—EXPLANATIONS

TASK G.7

1. Technician A says an enhanced evaporative emissions (EVAP) system must be able to detect a leak resulting from a 0.020 inch gap in the system. Technician B says the enhanced system must be able to detect a loose gas cap. Who is correct?

   A. A only

   B. B only

   C. Both A and B

   D. Neither A nor B

   Answer A is incorrect. Technician B is also correct.

   Answer B is incorrect. Technician A is also correct.

   Answer C is correct. Both Technicians are correct. An enhanced EVAP system must be able to detect leaks as small as 0.020 inch. A loose gas cap can be detected by an enhanced EVAP system.

   Answer D is incorrect. Both Technicians are correct.

2. An engine has a low power complaint. A power balance test has been performed, and all the cylinders have about the same amount power output. Which of the following could be the cause of the low power concern?

    A.   A restricted exhaust system

    B.   A shorted coil pack

    C.   A shorted spark plug wire

    D.   A burnt exhaust valve

TASK A.7

    Answer A is correct. A restricted exhaust will cause low power and will not be located during a power balance test.

    Answer B is incorrect. A shorted coil pack would most likely show up during a cylinder power balance test.

    Answer C is incorrect. A shorted spark plug wire would most likely show up during a cylinder power balance test.

    Answer D is incorrect. A burnt exhaust valve will cause a low power on one cylinder during a power balance test.

3. All of the following can be used to test a throttle position sensor (TPS) EXCEPT:

    A.   Oscilloscope

    B.   Ohmmeter

    C.   Ammeter

    D.   Scan tool

TASK H.6

    Answer A is incorrect. The voltage trace of a TPS can be tested with an oscilloscope to look for a momentarily open circuit.

    Answer B is incorrect. An ohmmeter can be used to measure the resistance of the sensor.

    Answer C is correct. There are no recognized tests using an ammeter on a TPS.

    Answer D is incorrect. The scan tool can be an effective means of monitoring the TPS.

4. The acronym CKP is used for:

    A.   Camshaft Position Sensor

    B.   Crankcase Pressure Sensor

    C.   Cylinder Pressure Sensor

    D.   Crankshaft Position Sensor

TASK E.6

    Answer A is incorrect. The camshaft position sensor acronym is CMP.

    Answer B is incorrect. There is no universal acronym for the crankcase pressure sensor.

    Answer C is incorrect. There is no universal acronym for the cylinder pressure sensor.

    Answer D is correct. CKP is the acronym for Crankshaft Position Sensor.

TASK F.7

5. Technician A says that prior to adjusting the throttle plates on a multi-port injected engine throttle body, the throttle plates and bore should be cleaned if dirty or varnished. Technician B says the TPS needs to be readjusted on some vehicles after the throttle plate angle adjustment is complete. Who is correct?

    A. A only
    B. B only
    C. Both A and B
    D. Neither A nor B

    Answer A is incorrect. Technician B is also correct.

    Answer B is incorrect. Technician A is also correct.

    Answer C is correct. Both Technicians are correct. If the throttle body is dirty or varnished, it will disturb the airflow into the engine. This can result in an improper throttle plate adjustment. If the throttle plates are adjusted, the angle has changed and the TPS needs to be adjusted as well.

    Answer D is incorrect. Both Technicians are correct.

TASK E.5

6. To check coil available voltage output, the technician should:

    A. Disconnect the fuel pump power lead
    B. Disconnect the plug wire at the plug and ground it
    C. Disconnect the coil wire and ground it
    D. Conduct the test using a suitable spark tester that requires 25 kV

    Answer A is incorrect. The fuel pump is not part of this test.

    Answer B is incorrect. Grounding the wire would result in a direct short.

    Answer C is incorrect. Grounding the wire would result in a direct short.

    Answer D is correct. An approved spark tester will require the coil to put out approximately 25 kV, usually sufficient for all engines.

TASK D.9,
D.10

7. The radiator is low on coolant, and the recovery bottle is overfull. Which of the following could be the cause?

    A. Tight water pump bearings
    B. Leaking radiator cap seal
    C. Incorrect cooling fan installed
    D. Incorrect fan clutch installed

    Answer A is incorrect. The coolant is not returning to the radiator when the engine cools. Tight water pump bearings would not cause this.

    Answer B is correct. A leaking radiator cap could allow air to be drawn back into the cooling system during cool down. This would keep the coolant in the recovery bottle and allow air into the radiator.

    Answer C is incorrect. If the incorrect fan were installed, the engine may overheat, but it would not cause the condition listed here.

    Answer D is incorrect. If the incorrect fan clutch were installed, the engine may overheat, but it would not cause the condition found in this question.

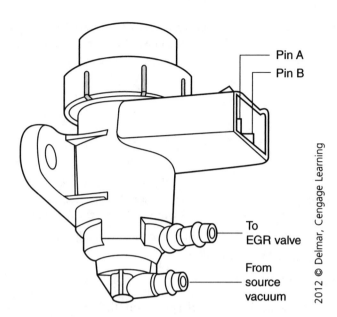

Pin A
Pin B

To
EGR valve

From
source
vacuum

2012 © Delmar, Cengage Learning

8. A diagnostic trouble code (DTC) for an exhaust gas recirculation (EGR) problem is retrieved. A reading of infinite between Pin A and Pin B of the EGR vacuum regulator valve shown in the above illustration could mean:

   A. Nothing
   B. The regulator is defective
   C. The EGR valve is defective
   D. The manifold absolute pressure (MAP) sensor is defective

TASK G.2

Answer A is incorrect. The windings should always have resistance.

Answer B is correct. This reading of infinite could mean that the regulator is defective. It has an open winding.

Answer C is incorrect. While a faulty EGR valve can cause a DTC for EGR, the vacuum regulator is what failed the test. The EGR valve is not being tested in this illustration.

Answer D is incorrect. A defective MAP sensor would not be located by measuring the resistance of the EGR vacuum regulator.

9. Technician A says a defective ECT sensor may cause a no-start condition. Technician B says a defective ECT sensor may cause the customer to have a transmission shift concern. Who is correct?

   A. A only
   B. B only
   C. Both A and B
   D. Neither A nor B

TASK H.6

Answer A is incorrect. Technician B is also correct.

Answer B is incorrect. Technician A is also correct.

Answer C is correct. Both Technicians are correct. The coolant temperature sensor input to the ECM is used to determine the correct amount of fuel needed by the engine as well as when to shift the transmission. Therefore, an incorrect signal could result in an engine that fails to start or a transmission that shifts too early or too late.

Answer D is incorrect. Both Technicians are correct.

TASK B.7

10. While adjusting valves with mechanical lifters, Technician A says when the valve clearance is checked on a cylinder, the piston should be positioned at top dead center (TDC) on the compression stroke. Technician B says the valves should be open when adjusted. Who is correct?

A. A only

B. B only

C. Both A and B

D. Neither A nor B

Answer A is correct. Only Technician A is correct. When the valve clearance is checked on a given cylinder, typically that cylinder will be at TDC compression. Some engine manufacturers have used a shim and bucket-style adjustment procedure for the valves.

Answer B is incorrect. The valves must be closed in order to perform an adjustment.

Answer C is incorrect. Only Technician A is correct.

Answer D is incorrect. Technician A is correct.

TASK A.9

11. While performing a cylinder leakage test, air is heard escaping from the adjacent cylinder. This could be caused by:

A. Failure to position the engine correctly

B. A leaking exhaust valve

C. A leaking head gasket

D. A leaking intake valve

Answer A is incorrect. When the technician incorrectly positions the engine for this test, it allows air to leak from the intake or exhaust valve

Answer B is incorrect. A leaking exhaust valve allows air to leak from the exhaust system during this test.

Answer C is correct. A leaking head gasket can allow air to enter the adjacent cylinder.

Answer D is incorrect. A leaking intake valve can allow air to leak into the intake during this test.

TASK H.1

12. While performing a scan test on an OBD-II certified vehicle, a DTC P1336 is retrieved. Technician A says that a first digit, P, means the code is a generic trouble code. Technician B says that a second digit, 1, means the code is a manufacturer specific code. Who is correct?

A. A only

B. B only

C. Both A and B

D. Neither A nor B

Answer A is incorrect. A first digit of P in an OBD-II DTC stands for *powertrain* and is the module that set the code.

Answer B is correct. Only Technician B is correct. An OBD-II code with 1 as the second digit means the code is a manufacturer specific code, not a generic code.

Answer C is incorrect. Only Technician B is correct.

Answer D is incorrect. Technician B is correct.

13. An engine produces a bottom-end knock when started. Which of the following could be the cause?

    A. A bent pushrod
    B. One or more collapsed lifters
    C. Lack of oil pressure to the valve train
    D. Worn main bearings

TASK C.9,
B.6

Answer A is incorrect. Bent pushrods do not make a knocking noise. They can make a light ticking noise in the upper end of the engine.

Answer B is incorrect. Collapsed lifters will not make a knocking noise. They can make a light ticking noise.

Answer C is incorrect. A lack of oil pressure to the valve train would cause tapping, not knocking.

Answer D is correct. A worn main bearing could create bottom-end knock when first started. A loose crankshaft main bearing produces a dull, steady knock, while a loose crankshaft thrust bearing produces a heavy thump at irregular intervals. The thrust bearing noise might only be audible on very hard acceleration. Both of these bearing noises are usually caused by worn bearings or crankshaft journals. To correct the problem, replace the bearings or crankshaft.

14. A truck with port fuel injection is running roughly. A lab scope shows each injector waveform to be identical, except for one that has a considerably shorter voltage spike than the others. Which of the following is the most likely cause?

    A. Faulty powertrain control module (PCM)
    B. Open connection at the injector
    C. Shorted injector winding
    D. Low charging system voltage

TASK F.8

Answer A is incorrect. A bad PCM would affect both injectors.

Answer B is incorrect. An open connection would result in the injector not working at all.

Answer C is correct. A shorted injector winding is the most likely cause of this condition. A shorted winding will result in a shorter voltage spike.

Answer D is incorrect. A low charging system voltage would affect both injectors.

15. An air pump is being replaced. Technician A says new air pumps will usually come with a new switching valve. Technician B says that the new air pump may not come with a new pulley. Who is correct?

    A. A only
    B. B only
    C. Both A and B
    D. Neither A nor B

TASK G.8

Answer A is incorrect. New air pumps do not usually come with a switching valve, which is usually purchased separately.

Answer B is correct. Only Technician B is correct. If the pump is belt-driven, it may not come with a pulley. The technician may have to remove the pulley from the old pump and install it on the new one.

Answer C is incorrect. Only Technician B is correct.

Answer D is incorrect. Technician B is correct.

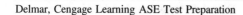

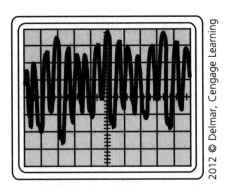

2012 © Delmar, Cengage Learning

TASK H.6

16. Based on the oxygen sensor waveform shown in the above illustration, all the following are true EXCEPT:

A. A lean biased condition is represented.

B. A DTC may be recorded in the PCM.

C. A rich biased condition is represented.

D. The oxygen sensor is functioning.

Answer A is correct. Oxygen sensor voltage is low under a lean condition and high under a rich condition. The voltage in this illustration is high. Therefore a rich condition is illustrated.

Answer B is incorrect. A condition as illustrated may set a rich biased DTC.

Answer C is incorrect. A high voltage reading on the oxygen sensor is represented.

Answer D is incorrect. This oxygen sensor is functioning. It is indicating a rich condition.

TASK A.7

17. Technician A says a power balance test can be performed with a scan tool on some trucks. Technician B says a power balance test can be performed using a test light on some trucks. Who is correct?

A. A only

B. B only

C. Both A and B

D. Neither A nor B

Answer A is incorrect. Technician B is also correct.

Answer B is incorrect. Technician A is also correct.

Answer C is correct. Both Technicians are correct. Some electronic control modules (ECMs) will perform a cylinder power balance test when commanded by a scan tool. Also, power balance can be tested using a spark plug wire adapter and a test light to remove the spark plugs from the cylinder one at a time.

Answer D is incorrect. Both Technicians are correct.

TASK G.6

18. While replacing an air-injection manifold, which of the following should be done?

A. Remove the exhaust manifold from the vehicle, then remove the air-injection manifold.

B. Remove the air pump from the vehicle, and then remove the air-injection manifold.

C. Use high-temperature silicone in place of the gasket.

D. Apply penetrating oil to the nuts before removal.

Answer A is incorrect. The exhaust manifold is removed from the vehicle after the air-injection manifold.

Answer B is incorrect. The air pump does not have to be removed in order to remove the air-injection manifold.

Answer C is incorrect. High-temperature silicone is not used in place of the gasket.

Answer D is correct. Penetrating oil should be applied to the nuts before removal.

19. Technician A says a high resistance connection at the alternator electrical connector could cause excessive heat buildup in the connector. Technician B says a connector that is too tight can cause heat distortion. Who is correct?

TASK A.17, A16

   A. A only
   B. B only
   C. Both A and B
   D. Neither A nor B

   Answer A is correct. Only Technician A is correct. A high resistance connection could cause excessive heat buildup in the connector.

   Answer B is incorrect. A good tight connection reduces resistive heat from generating.

   Answer C is incorrect. Only Technician A is correct.

   Answer D is incorrect. Technician A is correct.

20. The LEAST LIKELY condition caused by a defective MAP sensor would be:

   A. A rich or lean air/fuel ratio
   B. Engine surging
   C. Excess fuel consumption
   D. Excessive idle speeds

TASK F.11

   Answer A is incorrect. A MAP sensor is directly related to the air/fuel mixture.

   Answer B is incorrect. A defective MAP sensor can cause engine surging.

   Answer C is incorrect. If the MAP sensor is bad, it can cause a rich air/fuel mixture.

   Answer D is correct. A defective MAP sensor would be LEAST likely to cause excessive idle speeds.

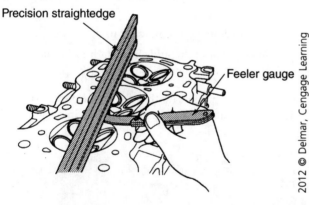

Precision straightedge

Feeler gauge

2012 © Delmar, Cengage Learning

21. In the above illustration, when measuring the cylinder head for warpage with a straightedge, the feeler gauge measurement is 0.012 in. (0.3 mm). Technician A says that measurement is higher than specified. Technician B says the measurement indicates that a thicker head gasket must be installed. Who is correct?

TASK B.3

   A. A only
   B. B only
   C. Both A and B
   D. Neither A nor B

   Answer A is correct. Only Technician A is correct. A typical maximum specification for cylinder head warpage is 0.003 inches..

   Answer B is incorrect. The cylinder head must be resurfaced or replaced.

   Answer C is incorrect. Only Technician A is correct.

   Answer D is incorrect. Technician A is correct.

22. A charging system fuse link is being replaced. Technician A says the new fuse link should be soldered in place. Technician B says the new fuse link should be one size larger than the original to prevent it from burning again. Who is correct?

    A. A only

    B. B only

    C. Both A and B

    D. Neither A nor B

Answer A is correct. Only Technician A is correct. Fuse links should always be soldered in place to help resist corrosion.

Answer B is incorrect. The replacement fuse link should be the same size as the original. If a larger size is installed, it is possible an electrical fire could result.

Answer C is incorrect. Only Technician A is correct.

Answer D is incorrect. Technician A is correct.

23. All of the following are true when testing a Magnetic Pulse Generator pickup coil EXCEPT:

    A. The pickup coil should be within the manufacturer's specified resistance value.

    B. An erratic reading while wiggling the pickup coil wires indicates that the pickup coil is intermittent.

    C. A resistance reading above manufacturer's specification indicates an open pickup coil.

    D. A resistance reading below manufacturer's specification indicates an open pickup coil.

Answer A is incorrect. The pickup coil should always be within specification.

Answer B is incorrect. Erratic readings while moving the pickup coil leads would indicate a bad connection.

Answer C is incorrect. A high resistance reading would indicate an open circuit.

Answer D is correct. An open pickup coil winding would show high or infinity on the ohmmeter. A reading below specification would indicate a shorted pick up coil.

24. A vehicle has an overheating concern. Technician A says that the water pump could be intermittently working. Technician B says that the thermostat could be intermittently sticking. Who is correct?

    A. A only

    B. B only

    C. Both A and B

    D. Neither A nor B

Answer A is incorrect. Water pumps do not work intermittently; they either pump or do not pump.

Answer B is correct. Only Technician B is correct. If the thermostat sometimes sticks, it could cause this condition.

Answer C is incorrect. Only Technician B is correct.

Answer D is incorrect. Technician B is correct.

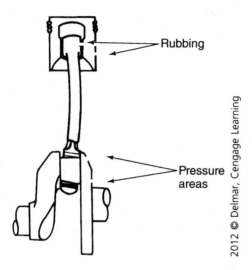

Rubbing

Pressure areas

2012 © Delmar, Cengage Learning

25. Referring to the above illustration, this condition may cause all of the following EXCEPT:

TASK C.3

    A. Uneven connecting-rod bearing wear

    B. Uneven main bearing wear

    C. Uneven piston pin wear

    D. Uneven cylinder wall wear

Answer A is incorrect. A bent connecting rod, as shown, can cause uneven wear on the connecting-rod bearing.

Answer B is correct. A bent connecting rod, as shown, will not usually cause wear on the main bearing.

Answer C is incorrect. A bent connecting rod, as shown, can cause wear on the piston pin.

Answer D is incorrect. A bent connecting rod, as shown, can cause a very specific cylinder wall wear pattern.

26. A low-pitched roaring noise is being diagnosed. The noise disappears when the accessory drive belt is removed. Technician A says the belt tensioner pulley bearing should be checked for roughness. Technician B says the alternator bearing should be checked for roughness. Who is correct?

TASK A.20

    A. A only

    B. B only

    C. Both A and B

    D. Neither A nor B

Answer A is incorrect. Technician B is also correct.

Answer B is incorrect. Technician A is also correct.

Answer C is correct. Both Technicians are correct. If a noise goes away when the belt is removed, the noise is in the accessory drive system. It could be the tensioner, alternator, or any other component that is being driven by the belt.

Answer D is incorrect. Both Technicians are correct.

TASK H.2

27. A truck has set a DTC PO171 (system too lean, bank 1). No other drivability concerns are present. The freeze frame data shows the code was set under warm idle conditions. Technician A says the problem could be an intake manifold vacuum leak. Technician B says the problem could be a weak fuel pump. Who is correct?

   A. A only

   B. B only

   C. Both A and B

   D. Neither A nor B

   Answer A is correct. Only Technician A is correct. A vacuum leak can set a fuel trim DTC with no other drivability problems present.

   Answer B is incorrect. A weak fuel pump would not cause problems at idle when fuel demand is low. Other drivability problems, such as lack of power, would be present.

   Answer C is incorrect. Only Technician A is correct.

   Answer D is incorrect. Technician A is correct.

TASK E.1

28. The LEAST LIKELY cause of spark knock is:

   A. EGR valve stuck closed

   B. Fuel quality

   C. Carbon buildup on top of the pistons

   D. EGR valve stuck open

   Answer A is incorrect. A stuck closed EGR valve can cause spark knock.

   Answer B is incorrect. Poor fuel quality can cause spark knock.

   Answer C is incorrect. Carbon buildup in the combustion chamber can cause spark knock.

   Answer D is correct. A stuck open EGR valve would not cause spark knock. Other drivability problems would be present.

TASK A.15

29. Technician A says the engine should be disabled to prevent engine start up during a starter current draw test. Technician B says that on some vehicles, the accelerator pedal can be held to the floor to prevent starting during a current draw test. Who is correct?

   A. A only

   B. B only

   C. Both A and B

   D. Neither A nor B

   Answer A is incorrect. Technician B is also correct.

   Answer B is incorrect. Technician A is also correct.

   Answer C is correct. Both Technicians are correct. When performing a starter current draw test, the engine must be disabled to prevent starting. Some vehicles have a clear flood mode that disables the injectors when the accelerator pedal is held to the floor.

   Answer D is incorrect. Both Technicians are correct.

30. Why are some fuel pressure regulators vacuum operated?

   A. To increase fuel delivery under high load conditions
   B. To prevent fuel-pressure leak down when the engine is turned off
   C. To provide a constant pressure drop across the injector due to changing manifold pressure
   D. To improve injector spray patterns

TASK F.6

Answer A is incorrect. Increased fuel delivery is provided through increased injector pulse width (PW).

Answer B is incorrect. Vacuum applied to the pressure regulator has no effect on leak down.

Answer C is correct. A vacuum-operated pressure regulator provides a constant pressure drop across the injector tip to compensate for changing intake manifold pressure.

Answer D is incorrect. Injector spray pattern is controlled by tip and orifice design.

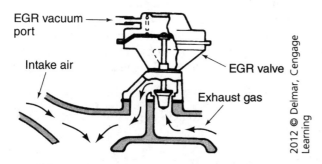

31. In the above illustration, Technician A says that EGR valves with heavy carbon buildup must be cleaned with a rotary file. Technician B says with the valve off the engine, fuel-injection cleaner may be used to clean the EGR valve's internal exhaust passages. Who is correct?

   A. A only
   B. B only
   C. Both A and B
   D. Neither A nor B

TASK G.5

Answer A is incorrect. If the carbon buildup on the EGR valve cannot be cleaned with solvent, the valve should be replaced. The rotary file will damage the valve.

Answer B is correct. Only Technician B is correct. Fuel-injection cleaner can be used to clean the internal passages in the EGR valve.

Answer C is incorrect. Only Technician B is correct.

Answer D is incorrect. Technician B is correct.

32. Technician A says that a valve with a margin measurement that is smaller than specified should be replaced. Technician B says that a valve with a stem measurement smaller than specified must be replaced. Who is correct?

   A. A only
   B. B only
   C. Both A and B
   D. Neither A nor B

TASK B.4

Answer A is incorrect. Technician B is also correct.

Answer B is incorrect. Technician A is also correct.

Answer C is correct. Both Technicians are correct. A valve with insufficient margin should be replaced, since it will likely fail soon and could cause other problems like hot spots in the combustion chamber. A valve with a stem that is smaller than specification is worn and must be replaced.

Answer D is incorrect. Both Technicians are correct.

33. Technician A says that enabling criteria are the specific conditions that must be met, such as ambient temperature or engine load, before a monitor will run. Technician B says that pending conditions are conditions that exist that prevent a specific monitor from running, such as an oxygen sensor fault code preventing a catalyst monitor from running. Who is correct?

    A. A only
    B. B only
    C. Both A and B
    D. Neither A nor B

    Answer A is incorrect. Technician B is also correct.

    Answer B is incorrect. Technician A is also correct.

    Answer C is correct. Both Technicians are correct. Enabling criteria are the specific operating parameters that must be met before a monitor will run. Many EVAP monitors, for instance, require a cold start with ambient temperature below a certain value. Pending conditions are any circumstances that may prevent a monitor from running properly, such as an oxygen sensor or coolant temperature sensor fault not allowing a catalyst monitor to run.

    Answer D is incorrect. Both Technicians are correct.

34. A four-gas exhaust emissions analyzer may be used to help diagnose all of the following problems EXCEPT:

    A. A slow or lazy oxygen sensor
    B. Engine cylinder misfire
    C. A defective catalytic converter
    D. A plugged or restricted fuel injector

    Answer A is correct. A digital storage oscilloscope is needed to properly test for slow or lazy oxygen sensors.

    Answer B is incorrect. A four-gas analyzer can indicate cylinder misfire.

    Answer C is incorrect. A four-gas analyzer can test catalytic converters.

    Answer D is incorrect. A four-gas analyzer can pinpoint a plugged fuel injector.

35. A vehicle stalls intermittently at idle and has low long-term fuel trim correction values stored when checked with a scan tool. All of the following conditions could cause this EXCEPT:

    A. Leaking fuel injectors
    B. Unmetered air leaking into the engine
    C. A leaking fuel-pressure regulator diaphragm
    D. A fuel-pressure regulator sticking closed

    Answer A is incorrect. Leaking fuel injectors could cause the symptoms listed.

    Answer B is correct. Unmetered air leaking into the engine would cause a lean air/fuel mixture and create high instead of low long-term fuel trim corrections.

    Answer C is incorrect. A leaking fuel-pressure regulator diaphragm could cause the symptoms listed.

    Answer D is incorrect. A sticking fuel-pressure regulator could cause the symptoms listed.

36. A truck comes into the shop with an oil pressure light flickering at idle. Technician A says the truck needs a new oil pump to correct the problem. Technician B says a known good oil pressure gauge should be installed on the engine to properly verify oil pressure. Who is correct?

TASK D.1, A.12

A. A only

B. B only

C. Both A and B

D. Neither A nor B

Answer A is incorrect. The pump is not necessarily faulty.

Answer B is correct. Only Technician B is correct. Never trust the vehicle's dash gauge for diagnostic purposes. Always check the oil pressure with a known good oil pressure gauge before trying to correct the problem.

Answer C is incorrect. Only Technician B is correct.

Answer D is incorrect. Technician B is correct.

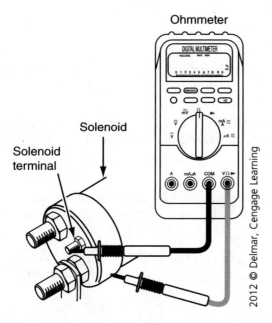

37. What procedure is being performed in the above illustration?

TASK A.17

A. Starter solenoid resistance test

B. Ignition current draw test

C. Battery load test

D. Starter solenoid available voltage test

Answer A is correct. An ohmmeter is being used to check the solenoid circuit resistance.

Answer B is incorrect. An ohmmeter is not used to check current draw.

Answer C is incorrect. A solenoid is shown, and an ohmmeter is not used to load test a battery.

Answer D is incorrect. Available voltage is tested with a voltmeter, not an ohmmeter.

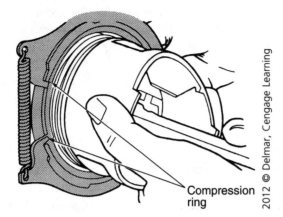

2012 © Delmar, Cengage Learning

Compression ring

TASK C.13

38. The tool in the above illustration is being used for what purpose?

   A. To widen the piston ring grooves
   B. To deepen the piston ring grooves
   C. To remove and replace the piston rings
   D. To remove carbon from the piston ring grooves

   Answer A is incorrect. The tool shown in the illustration is not used to widen piston ring grooves. In the past, the technician may have widened the groove and installed a spacer, but this is no longer a cost-effective repair technique.

   Answer B is incorrect. The tool shown in the illustration is not used to deepen piston ring grooves. Piston ring grooves are not deepened because this would weaken the piston.

   Answer C is correct. The tool shown in the illustration is used to remove and replace the piston rings.

   Answer D is incorrect. The tool shown in the illustration is not used to remove carbon from the piston ring grooves. A carbon-removing tool is shape more like a "C" with a cutting wheel installed.

TASK E.1

39. An engine equipped with a distributorless ignition system (DIS) will not start. Technician A says a defective CKP could cause this. Technician B says an open ground wire to the DIS assembly could be the cause. Who is correct?

   A. A only
   B. B only
   C. Both A and B
   D. Neither A nor B

   Answer A is incorrect. Technician B is also correct.

   Answer B is incorrect. Technician A is also correct.

   Answer C is correct. Both Technicians are correct. Either a failed CKP or an open ground connection to the DIS assembly would prevent proper primary circuit operation and cause a no-start condition.

   Answer D is incorrect. Both Technicians are correct.

40. Which of the following would most likely cause a single cylinder misfire?

    A. 5 percent cylinder leakage
    B. 50 percent cylinder Leakage
    C. A shorted CKP
    D. Retarded ignition timing

TASK A.8

Answer A is incorrect. 5 percent leakage is acceptable and will not cause a misfire.

Answer B is correct. 50 percent leakage is unacceptable and can cause a misfire.

Answer C is incorrect. A shorted CKP would likely affect all cylinders.

Answer D is incorrect. Retarded timing will not cause a misfire. It would affect all cylinders if the engine could stay running with the timing retarded far enough to cause misfire.

41. Technician A says that low cranking speed can be caused by high electrical resistance in the starter electrical circuit. Technician B says that a spun rod bearing can be the cause of the starter motor not being able to turn the engine over. Who is correct?

    A. A only
    B. B only
    C. Both A and B
    D. Neither A nor B

TASK A.18

Answer A is incorrect. Technician B is also correct.

Answer B is incorrect. Technician A is also correct.

Answer C is correct. Both Technicians are correct. High resistance in the starter electrical circuit can cause low current flow and therefore low cranking speed. A condition such as a spun rod bearing can cause the engine to be locked and can be the cause for the starter motor not turning the engine over.

Answer D is incorrect. Both Technicians are correct.

42. Technician A says that an EGR position sensor may be tested with either a voltmeter or an oscilloscope. Technician B says that if the PCM does not set an EGR position sensor fault code, there is no problem with the sensor. Who is correct?

    A. A only
    B. B only
    C. Both A and B
    D. Neither A nor B

TASK H.6

Answer A is correct. Only Technician A is correct. The EGR position sensor can be checked with a voltmeter or an oscilloscope.

Answer B is incorrect. If the PCM does not set an EGR position sensor fault code, there may still be a problem with the sensor.

Answer C is incorrect. Only Technician A is correct.

Answer D is incorrect. Technician A is correct.

TASK E.1

43. A primary ignition circuit on a vehicle checks good, but there is no spark from the coil wire. This could be caused by:

 A. A defective coil
 B. A grounded rotor
 C. An overheated transistor
 D. An open diode

 Answer A is correct. A defective coil could cause this condition.

 Answer B is incorrect. A rotor is located after a coil wire in the circuit. A grounded rotor would cause a no-start condition.

 Answer C is incorrect. The primary ignition circuit checks as good. An overheated transistor would cause the primary ignition circuit to fail and a no-start condition.

 Answer D is incorrect. A diode problem would not cause this failure. An open diode could cause a primary circuit failure and a no-start condition.

TASK A.5

44. Blue smoke emission from the tailpipe can be caused by all of the following EXCEPT:

 A. Worn valve guides
 B. Worn valve seals
 C. Fouled spark plugs
 D. Worn piston rings

 Answer A is incorrect. Worn valve guides make blue smoke.

 Answer B is incorrect. Worn valve seals also make blue smoke.

 Answer C is correct. A fouled spark plug will not allow the fuel to burn, resulting in black smoke. It will not cause blue smoke.

 Answer D is incorrect. Worm piston rings will make blue smoke.

TASK C.13

45. Technician A says that if the piston ring end gap is less than specified, a large piston ring must be used. Technician B says less than specified piston ring end gap can cause a piston ring to break. Who is correct?

 A. A only
 B. B only
 C. Both A and B
 D. Neither A nor B

 Answer A is incorrect. If the piston ring end gap is less than specified, the ring can usually be filed.

 Answer B is correct. Only Technician B is correct. Less than specified piston ring end gap can cause a piston ring to break. As the ring warms, it expands. The ends will butt together and crack the ring.

 Answer C is incorrect. Only Technician B is correct.

 Answer D is incorrect. Technician B is correct.

46. A truck with an electric fuel pump must be cranked for an excessive time before the engine will start. Technician A says a fuel leak down test should be performed. Technician B says there may be a fault in the fuel pump relay circuit. Who is correct?

TASK F.1

    A. A only

    B. B only

    C. Both A and B

    D. Neither A nor B

    Answer A is incorrect. Technician B is also correct.

    Answer B is incorrect. Technician A is also correct.

    Answer C is correct. Both Technicians are correct. A fuel leak down test should be performed to see if residual pressure bleeds off too rapidly due to internal or external leaks. This causes excessive cranking times while the system is re-pressurized. There may be a fault in the fuel pump relay circuit. When the fuel pump relay fails to supply voltage to the fuel pump, some vehicles use a back-up circuit using the oil pressure sender. If the engine is cranked long enough for oil pressure to build up sufficiently enough to close the contacts in this switch, the engine will start.

    Answer D in incorrect. Both Technicians are correct.

47. When performing a cylinder leakage test, which of the following is usually considered to be the maximum acceptable percentage?

TASK A.9

    A. 5 to 10 percent

    B. 15 to 20 percent

    C. 30 to 40 percent

    D. 40 to 45 percent

    Answer A is incorrect. While 5 to 10 percent is an acceptable cylinder leakage test result, it is too low to be considered the maximum acceptable amount. .

    Answer B is correct. 15 to 20 percent is considered acceptable. Most engines, even new ones, experience some leakage around the rings. Up to 20 percent is considered acceptable during the leakage test. When the engine is actually running, the rings will seal much better and the actual percent of leakage will be lower. However, no leakage should occur around the valves or the head gasket.

    Answer C is incorrect. 30 to 40 percent is too high. This much leakage can cause high crankcase pressure.

    Answer D is incorrect. 40 to 45 percent is too high. This much leakage will cause a low power concern, and a possible cylinder misfire.

48. A truck is brought to the shop with a fuel-related DTC. Technician A says all codes should be recorded and then erased to see if any are active codes. Technician B says some scan tools list active and inactive for ease of interpretation. Who is correct?

TASK H.2

    A. A only

    B. B only

    C. Both A and B

    D. Neither A nor B

    Answer A is incorrect. Technician B is also correct.

    Answer B is incorrect. Technician A is also correct.

    Answer C is correct. Both Technicians are correct. Some scan tools do sort active and inactive codes; however, it is good practice to record DTCs, erase them from the PCM, and then drive the vehicle to see if any reoccur.

    Answer D is incorrect. Both Technicians are correct.

TASK B.4

49. Engine valve springs should be inspected for all of the following EXCEPT:

    A. Cracks
    B. Spring tension
    C. Temperature resistance
    D. Spring squareness

    Answer A is incorrect. Springs should be inspected for cracks.

    Answer B is incorrect. Valve springs' tension should be inspected.

    Answer C is correct. Temperature resistance is not checked.

    Answer D is incorrect. The spring should be inspected to ensure it is square.

**TASK A.15,
C.10**

50. While performing a starter current draw test on a truck, a technician records 75 amps. Technician A says this indicates excessive resistance in the starter control circuit. Technician B says 75 amps is an excessive current draw. Who is correct?

    A. A only
    B. B only
    C. Both A and B
    D. Neither A nor B

    Answer A is incorrect. Technician A is incorrect.

    Answer B is incorrect. Technician B is incorrect.

    Answer C is incorrect. Neither Technician is correct.

    Answer D is correct. Neither Technician is correct. 75 amps is too small of a starter current draw. This reading may be caused by high electrical resistance or by an engine that does not have enough mechanical internal resistance. An example of this latter condition may be a jumped timing chain.

PREPARATION EXAM 5—ANSWER KEY

1.	B	21.	C	41.	B
2.	D	22.	D	42.	C
3.	A	23.	A	43.	B
4.	A	24.	D	44.	A
5.	D	25.	C	45.	C
6.	C	26.	B	46.	C
7.	C	27.	C	47.	B
8.	A	28.	C	48.	C
9.	D	29.	B	49.	B
10.	C	30.	C	50.	C
11.	A	31.	D		
12.	B	32.	C		
13.	D	33.	B		
14.	C	34.	C		
15.	C	35.	D		
16.	A	36.	C		
17.	C	37.	D		
18.	D	38.	C		
19.	D	39.	B		
20.	C	40.	D		

PREPARATION EXAM 5—EXPLANATIONS

1. Technician A says valve adjustment should always be performed on a running engine. Technician B says the piston should be placed at top dead center (TDC) of the compression stroke. Who is correct?

   TASK B.7

   A. A only

   B. B only

   C. Both A and B

   D. Neither A nor B

   Answer A is incorrect. Most manufacturers specify that the engine not be running.

   Answer B is correct. Only Technician B is correct. Many adjustment procedures require the piston at TDC on the compression stroke as the intake and exhaust valves are closed.

   Answer C is incorrect. Only Technician B is correct.

   Answer D is incorrect. Technician B is correct.

TASK E.7

2. The ignition control module uses the digital signal received from the powertrain control module (PCM) for:

   A.   RPM input

   B.   Hall effect timing

   C.   Cylinder #1 signal

   D.   Computed timing signal

Answer A is incorrect. The ignition module does not need an RPM signal.

Answer B is incorrect. There is no such thing as Hall effect timing.

Answer C is incorrect. The ignition module does not discriminate between cylinders.

Answer D is correct. The signal from the PCM is the result of computed inputs for proper timing advance. This is the signal the ignition module uses to fire the coil at the proper time.

TASK F.7

3. Technician A says a faulty throttle position sensor (TPS) can cause a hesitation on acceleration. Technician B says a faulty TPS will always set a diagnostic trouble code (DTC). Who is correct?

   A.   A only

   B.   B only

   C.   Both A and B

   D.   Neither A nor B

Answer A is correct. Only Technician A is correct. Incorrect inputs can result in performance or emission problems.

Answer B is incorrect. A faulty TPS sensor does not always cause a DTC. A midrange failure may not cause values to drop outside set parameters.

Answer C is incorrect. Only Technician A is correct.

Answer D is incorrect. Technician A is correct.

TASK F.3

4. The LEAST LIKELY cause of a fuel tank leak is:

   A.   Defective tank straps

   B.   Road damage

   C.   Defective seams

   D.   Corrosion

Answer A is correct. Defective tank straps would be the least likely cause of a fuel tank leak.

Answer B is incorrect. Road damage often causes fuel tank damage.

Answer C is incorrect. Defective seams can be a cause of fuel leaks.

Answer D is incorrect. Corrosion, after it has worn through the tank, will cause a fuel leak.

5. When replacing a programmable read only memory (PROM), Technician A says that you should never ground yourself to the vehicle. Technician B says that grounding yourself to the vehicle will erase the PROM. Who is correct?

    A. A only

    B. B only

    C. Both A and B

    D. Neither A nor B

    TASK H.9

    Answer A is incorrect. When replacing a PROM, always ground yourself to the vehicle to reduce static discharge.

    Answer B is incorrect. Being grounded will not erase the PROM.

    Answer C is incorrect. Neither Technician is correct.

    Answer D is correct. Neither Technician is correct. When replacing a PROM, always ground yourself to the vehicle to reduce static discharge. Being grounded will not erase the PROM. A Static discharge of electricity can erase or otherwise damage the PROM.

6. A slipped timing belt can cause all of the following EXCEPT:

    A. Poor fuel mileage

    B. A no-start condition

    C. High manifold vacuum

    D. Low power

    TASK C.5

    Answer A is incorrect. Poor fuel mileage is a result of improper timing.

    Answer B is incorrect. If the belt has slipped enough, the engine will not start.

    Answer C is correct. Vacuum readings would tend to be lower than normal.

    Answer D is incorrect. Improper timing can cause a lack of power.

7. Technician A says a vacuum leak decreases engine performance. Technician B says propane is a good method of locating vacuum leaks. Who is correct?

    A. A only

    B. B only

    C. Both A and B

    D. Neither A nor B

    TASK A.6

    Answer A is incorrect. Technician B is also correct.

    Answer B is incorrect. Technician A is also correct.

    Answer C is correct. Both Technicians are correct. Any vacuum leak will change the air/fuel mixture, resulting in loss of performance and fuel economy. Controlled use of propane gas can help confirm and locate the source of vacuum leaks.

    Answer D is incorrect. Both Technicians are correct.

8. A computer ground circuit is being tested for voltage drop. Which reading below would be acceptable?

    A. 0.05 V

    B. 0.9 V

    C. 1.1 V

    D. 1.9 V

    TASK H.7

    Answer A is correct. This would be an acceptable voltage drop for a computer ground.

    Answer B is incorrect. This is too high. It could cause computer problems.

    Answer C is incorrect. This is too high. It could cause computer problems.

    Answer D is incorrect. This is much too high. This level of voltage drop would definitely cause sensor-related problems.

TASK E.6

9. A magnetic pulse crank sensor is being tested with an ohmmeter. Technician A says when the pickup coil leads are moved, an erratic ohmmeter reading is normal. Technician B says that an infinite ohmmeter reading between the pickup coil terminals is an acceptable reading. Who is correct?

A. A only

B. B only

C. Both A and B

D. Neither A nor B

Answer A is incorrect. If the pickup coil leads are moved, there should be no erratic reading on the ohmmeter. If there is, look for damaged pickup coil wires.

Answer B is incorrect. An infinite reading indicates an open circuit.

Answer C is incorrect. Neither Technician is correct.

Answer D is correct. Neither Technician is correct. If the pickup coil leads are moved, there should be no erratic reading on the ohmmeter. If there is, look for damaged pickup coil wires. An infinite reading indicates an open circuit. A normal reading would be about 1000 ohms.

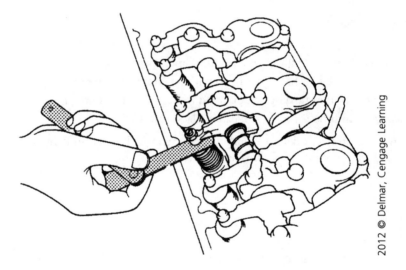

2012 © Delmar, Cengage Learning

TASK B.7

10. Technician A says that if the measurement in the above illustration is set too wide, it will retard valve timing. Technician B says it will reduce valve overlap. Who is correct?

A. A only

B. B only

C. Both A and B

D. Neither A nor B

Answer A is incorrect. Technician B is also correct.

Answer B is incorrect. Technician A is also correct.

Answer C is correct. Both Technicians are correct. The valve will open late and close early, reducing flow and overlap.

Answer D is incorrect. Both Technicians are correct.

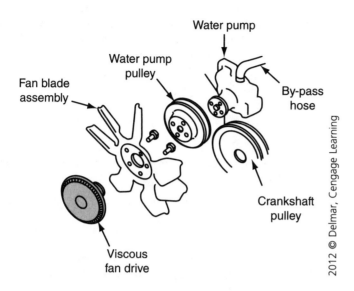

2012 © Delmar, Cengage Learning

Water pump

Water pump
pulley

Fan blade
assembly

By-pass
hose

Crankshaft
pulley

Viscous
fan drive

11. A technician is testing a viscous-drive fan clutch, as shown in the above illustration, with the
    engine off. When the technician rotates the cooling fan by hand, it should have:

    A. More resistance hot
    B. More resistance cold
    C. No rotation movement
    D. No resistance

TASK D.11

Answer A is correct. Hotter temperatures result in more resistance, thus causing the fan to
move more air across the radiator.

Answer B is incorrect. It has less resistance cold, when fan-blown air is not needed to
improve performance and fuel economy.

Answer C is incorrect. It should have movement.

Answer D is incorrect. The clutch should have resistance.

12. A vehicle is being diagnosed for a drivability complaint. The technician finds a soft code
    stored in the PCM. Technician A says the soft code is a code that would be classified as a
    C-type code. Technician B says the soft code is a code that occurred in the past and no longer
    exists. Who is correct?

TASK H.2

    A. A only
    B. B only
    C. Both A and B
    D. Neither A nor B

Answer A is incorrect. A soft code is not a C-type code. A C-code is a chassis system code.

Answer B is correct. Only Technician B is correct. A soft code is a code that occurred in the
past and no longer exists.

Answer C is incorrect. Only Technician B is correct.

Answer D is incorrect. Technician B is correct.

TASK G.2

13. Technician A says that with the PCV valve disconnected from the rocker cover, there should be no vacuum at the valve with the engine idling. Technician B says that when the PCV valve is removed and shaken, there should not be a rattling noise. Who is correct?

   A. A only
   B. B only
   C. Both A and B
   D. Neither A nor B

   Answer A is incorrect. Since the valve is located downstream from the throttle plates, there will always be a vacuum at the PCV valve.

   Answer B is incorrect. There should be a noise when the valve is shaken, proving the plunger is not stuck.

   Answer C is incorrect. Neither Technician is correct.

   Answer D is correct. Neither Technician is correct. A plugged vacuum line can cause there to be no vacuum at the valve. If the valve does not rattle, the valve could be missing pieces.

TASK A.9

14. All of the following are true of the cylinder leakage test EXCEPT:

   A. Air loss and bubbles in the radiator indicate a bad head gasket or engine casting crack.
   B. Air loss from the oil filler cap indicates worn piston rings.
   C. A gauge reading of 100 percent indicates no cylinder leakage.
   D. Air loss from the exhaust indicates a valve problem.

   Answer A is incorrect. Air loss and bubbles in the radiator indicate a blown head gasket or cracked head or block casting.

   Answer B is incorrect. Air escaping from the crankcase could indicate problems with the rings.

   Answer C is correct. A reading of 100 percent indicates a major, total cylinder leak, either due to an incorrect crank position (valve open) or internal damage. A reading of up to 20 percent is considered normal.

   Answer D is incorrect. Air escaping from the exhaust indicates a bad exhaust valve.

TASK A.14

15. Technician A says a key-off current draw on a battery of 0.05 amps is acceptable for most cars. Technician B says that under most operating conditions, the alternator supplies the current for vehicle electrical loads once the engine is running. Who is correct?

   A. A only
   B. B only
   C. Both A and B
   D. Neither A nor B

   Answer A is incorrect. Technician B is also correct.

   Answer B is incorrect. Technician A is also correct.

   Answer C is correct. Both Technicians are correct. Key-off current draw is called a parasitic load and should be less than 0.05 amps. The alternator supplies the vehicle's electrical current when the engine is running under all normal driving conditions. During engine idle with high electrical load, the current requirements may exceed alternator output until engine speed increases. The battery supplies the additional current needed.

   Answer D is incorrect. Both Technicians are correct.

16. Which of these would cause a double knocking noise with the engine at an idle?

    A. Worn piston wrist pins
    B. Excessive timing chain deflection
    C. Worn main bearing
    D. Excessive main bearing thrust clearance

TASK C.11

    Answer A is correct. Worn piston wrist pins create a double knock because the piston changes direction at TDC and bottom dead center (BDC) every crankshaft revolution.

    Answer B is incorrect. Excessive timing chain deflection can cause a rattle or a low vacuum, but it will not create a double knock.

    Answer C is incorrect. A worn main bearing will cause a single knock, not a double knock.

    Answer D is incorrect. Excessive main bearing thrust clearance can cause excessive crankshaft end-play, but it will not create a double knock.

17. An engine has a lack of power and excessive fuel consumption. Technician A says a broken timing belt cannot be the cause. Technician B says the timing belt may have jumped a tooth. Who is correct?

    A. A only
    B. B only
    C. Both A and B
    D. Neither A nor B

TASK B.8

    Answer A is incorrect. Technician B is also correct.

    Answer B is incorrect. Technician A is also correct.

    Answer C is correct. Both Technicians are correct. If the timing belt were broken, the engine would not run. A jumped tooth on the timing belt would change valve timing and volumetric efficiency.

    Answer D is incorrect. Both Technicians are correct.

18. When diagnosing fuel-injection system problems, a technical service bulletin search is performed for all of the following reasons EXCEPT:

    A. To save diagnostic time
    B. To locate service manual updates or specification changes
    C. Mid-year production changes
    D. Year, make, and model identification

TASK A.4

    Answer A is incorrect. Researching service bulletins can save diagnostic time.

    Answer B is incorrect. Changes to specifications are found in service bulletins.

    Answer C is incorrect. Service bulletins alert technicians to production changes made during the model year.

    Answer D is correct. Year, make, and model identification must already be known in order to look up information in the service bulletin.

TASK E.3

19. A test lamp is connected between the negative side of the coil and ground to diagnose a no-start condition. Technician A says a flickering test lamp could be caused by a defective ignition module. Technician B says a flickering test lamp could be caused by a defective pickup coil. Who is correct?

A. A only

B. B only

C. Both A and B

D. Neither A nor B

Answer A is incorrect. A flickering test lamp indicates that the module is working properly.

Answer B is incorrect. A flickering test lamp indicates the pickup coil is working properly.

Answer C is incorrect. Neither Technician is correct.

Answer D is correct. Neither Technician is correct. If the test lamp flickers, primary switching is occurring, indicating the pickup and module are functioning. If the light does not flicker, it indicates that the primary side of the ignition system is not opening and closing.

TASK H.9

20. All of the following are part of PCM reprogramming EXCEPT:

A. Connect battery maintainer.

B. Check technical service bulletins (TSBs) for current updates.

C. Disconnect the negative battery cable.

D. Turn ignition key to the on position.

Answer A is incorrect. Connecting a battery maintainer is part of reprogramming the PCM.

Answer B is incorrect. Always check for current TSBs before reprogramming a PCM.

Answer C is correct. You must have full battery power when reprogramming. Both cables should have a good electrical connection.

Answer D is incorrect. The key must be turned on to power up the PCM.

TASK F.4

21. Technician A says a fuel-pressure test will test the ability of the fuel pump to provide pressure. Technician B says it is possible for a fuel pump to pass a fuel-pressure test and still fail to provide sufficient flow. Who is correct?

A. A only

B. B only

C. Both A and B

D. Neither A nor B

Answer A is incorrect. Technician B is also correct.

Answer B is incorrect. Technician A is also correct.

Answer C is correct. Both Technicians are correct. The pressure test is done to prove pump operation, thus confirming power, ground, and pump action. Specifications are published for fuel pressure and volume. A fuel pump can be capable of providing pressure. However, a restricted fuel filter could prevent the pump from providing sufficient flow.

Answer D is incorrect. Both Technicians are correct.

22.　A vehicle has a rough idle with black smoke coming out of the exhaust. The engine smoothes out when the exhaust gas recirculation (EGR) valve is lightly tapped with a hammer. Technician A says the EGR valve vacuum hose could be leaking. Technician B says the EGR valve pressure differential sensor could be faulty. Who is correct?

TASK G.3

　　A.　A only

　　B.　B only

　　C.　Both A and B

　　D.　Neither A nor B

Answer A is incorrect. When an EGR valve sticks open, sometimes lightly tapping at the base of the valve will free it temporarily. If the EGR vacuum hose is leaking, the most likely result is a valve that fails to open or will not open the required amount.

Answer B is incorrect. Tapping the valve would not make a change in the pressure differential sensor.

Answer C is incorrect. Neither Technician is correct.

Answer D is correct. Neither Technician is correct. The most likely cause is a sticking EGR valve. This valve would most likely be replaced.

23.　The LEAST LIKELY cause of blue exhaust smoke is:

TASK A.5

　　A.　Worn valve seats

　　B.　Worn valve seals

　　C.　Stuck piston rings

　　D.　Worn cylinder walls

Answer A is correct. Worn valve seats cause compression loss, not oil consumption.

Answer B is incorrect. Worn valve seals are a common cause of blue exhaust smoke.

Answer C is incorrect. Stuck piston rings will cause oil consumption and smoke.

Answer D is incorrect. Worn or tapered cylinder walls increase oil burning and smoke.

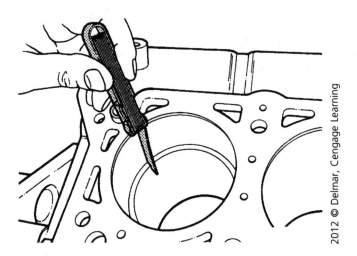

2012 © Delmar, Cengage Learning

TASK C.11

24. When measuring the piston ring end gap, as shown in the above illustration, Technician A says that the ring gap should be measured with the ring positioned at the middle of the ring travel in the cylinder. Technician B says that ring gap should not exceed 0.004 inches in a 4-inch diameter bore. Who is correct?

A. A only

B. B only

C. Both A and B

D. Neither A nor B

Answer A is incorrect. Ring end gap is measured at the bottom of the cylinder. This is where the least end clearance will be found because this is where the smallest bore diameter is located.

Answer B is incorrect. Ring end gap is usually about 0.004 for each inch of cylinder diameter.

Answer C is incorrect. Neither Technician is correct.

Answer D is correct. Neither Technician is correct. Ring end gap is measured at the bottom of the cylinder, not the top. This ensures that the minimum measurement is taken. Ring end gap is usually about 0.003 in–0.004 in. for each inch of cylinder diameter, so a 4-inch diameter cylinder would have a specification of about 0.016 in. A ring end gap of 0.004 inches for this cylinder is too small.

TASK E.7

25. Technician A says the ignition module may control ignition coil dwell time on some ignition systems. Technician B says the ignition module may control ignition coil current flow on some ignition systems. Who is correct?

A. A only

B. B only

C. Both A and B

D. Neither A nor B

Answer A is incorrect. Technician B is also correct.

Answer B is incorrect. Technician A is also correct.

Answer C is correct. Both Technicians are correct. The ignition module on electronic ignition systems can control ignition coil dwell time or current flow.

Answer D is incorrect. Both Technicians are correct.

26. While monitoring secondary ignition with an oscilloscope, the LEAST LIKELY cause of high resistance in the ignition secondary circuit is:

    A. Damaged spark plug wires
    B. No dielectric compound on the ignition module mounting surface
    C. Corroded spark plug wire ends
    D. Excessive spark plug air gap

    TASK E.4

    Answer A is incorrect. Damaged carbon ignition wires cause high resistance.

    Answer B is correct. The ignition module is common to all cylinders and is part of the primary ignition circuit.

    Answer C is incorrect. Corroded spark plug wire ends will cause high resistance.

    Answer D is incorrect. An excessive spark plug gap causes high resistance.

27. A vehicle with a MIL lamp illuminated is being diagnosed. The DTC stored in the PCM is PO303 (cylinder #3 misfire). Technician A says this is a one-trip failure because catalyst damage can occur. Technician B says the MIL will flash when this code is stored. Who is correct?

    A. A only
    B. B only
    C. Both A and B
    D. Neither A nor B

    TASK H.2

    Answer A is incorrect. Technician B is also correct.

    Answer B is incorrect. Technician A is also correct.

    Answer C is correct. Both Technicians are correct. Cylinder misfires are one-trip failures that cause the MIL lamp to flash.

    Answer D is incorrect. Both Technicians are correct.

28. A vacuum gauge is connected to an engine. When the engine is accelerated and held at a steady speed, the gauge slowly drops to 8" vacuum. Which of the following could be the problem?

    A. Sticking valves
    B. Over-advanced ignition timing
    C. A restricted exhaust
    D. A rich fuel mixture

    TASK A.6

    Answer A is incorrect. Sticking valves would cause the needle on the vacuum gauge to be jumpy, not falling.

    Answer B is incorrect. Advanced ignition timing would not cause vacuum to decrease. It could cause the needle to be higher than normal or jumpy if the timing is so far advanced it is causing a misfire.

    Answer C is correct. An exhaust restriction reduces airflow through the engine and will increase intake manifold pressure, which reduces the vacuum reading.

    Answer D is incorrect. Rich fuel mixtures would not cause a vacuum decrease off idle. Rich mixtures can cause a floating needle if the mixture is so rich it is causing a misfire.

29. Technician A says the electrolyte level is not important in a non-serviceable battery. Technician B says that on some batteries, the electrolyte level can be checked in a sealed battery by looking through the translucent battery case. Who is correct?

    A. A only

    B. B only

    C. Both A and B

    D. Neither A nor B

Answer A is incorrect. The level of the electrolyte is critical in any battery, regardless of its serviceability.

Answer B is correct. Only Technician B is correct. On some batteries, the case is translucent with the full line clearly marked, which enables the technician to check the electrolyte level.

Answer C is incorrect. Only Technician B is correct.

Answer D is incorrect. Technician B is correct.

30. Technician A says some evaporative emissions canisters have a replaceable filter. Technician B says if the filler cap is equipped with pressure and vacuum valves, they must be checked for dirt contamination and damage. Who is correct?

    A. A only

    B. B only

    C. Both A and B

    D. Neither A nor B

Answer A is incorrect. Technician B is also correct.

Answer B is incorrect. Technician A is also correct.

Answer C is correct. Both Technicians are correct. Some canisters have a replaceable filter. A restricted canister filter can prevent fresh air flow through the canister, causing ineffective purging. It should be checked or replaced at recommended service intervals. Gas caps can be checked for pressure and vacuum holds. There are equipment and specifications available for this test.

Answer D is incorrect. Both Technicians are correct.

31. A technician is performing a compression test. Which statement below is LEAST LIKELY to be true?

    A. Higher than normal readings on all cylinders could be caused by carbon buildup.

    B. Even but lower than normal readings on all cylinders could be caused by a slipped timing chain.

    C. Low readings on two adjacent cylinders might be caused by a blown head gasket.

    D. A low reading on one cylinder may be caused by a vacuum leak at that cylinder.

Answer A is incorrect. Carbon buildup could cause high readings.

Answer B is incorrect. Consistent low readings could be caused by a slipped timing chain.

Answer C is incorrect. Low readings on adjacent cylinders usually indicate a blown head gasket.

Answer D is correct. A vacuum leak affects air/fuel mixture, not cylinder sealing.

32. A vehicle needs to have its PCM re-flashed. Technician A says the vehicle should be connected to a battery maintainer to prevent accidental battery discharge during flashing. Technician B says the scan tool connector must fit snugly in the diagnostic link connector (DLC). Who is correct?

TASK H.9

A. A only

B. B only

C. Both A and B

D. Neither A nor B

Answer A is incorrect. Technician B is also correct.

Answer B is incorrect. Technician A is also correct.

Answer C is correct. Both Technicians are correct. When re-flashing a PCM, a battery maintainer should be connected to ensure the battery stays fully charged during the re-flashing, and the scan tool must fit the DLC snugly. A loose battery or scan tool connection can cause a total loss of data.

Answer D is incorrect. Both Technicians are correct.

33. While performing a valve adjustment, Technician A says the crankshaft must be placed in the proper position so that the valve lifter is riding on the camshaft lobe. Technician B says that adjusting valves with too little clearance may cause rough running and burnt valves. Who is correct?

TASK B.7

A. A only

B. B only

C. Both A and B

D. Neither A nor B

Answer A is incorrect. The crankshaft must be in the proper position so that the valve lifter is on the base circle of the camshaft, not the cam lobe.

Answer B is correct. Only Technician B is correct. Setting valve clearance too tight can cause the valve to not seat when the engine is warm, and combustion pressure will leak past the valve. Rough idle and valve burning will result.

Answer C is incorrect. Only Technician B is correct.

Answer D is incorrect. Technician B is correct.

34. While testing the cooling system, Technician A says to repeat the pressure test after repairs are made to ensure that all leaks are found. Technician B says a pressure test should include testing the radiator cap. Who is correct?

TASK D.6

A. A only

B. B only

C. Both A and B

D. Neither A nor B

Answer A is incorrect. Technician B is also correct.

Answer B is incorrect. Technician A is also correct.

Answer C is correct. Both Technicians are correct. The cooling system may have more than one leak. The cap must be tested to confirm it will hold pressure. There is a specification for the system pressure.

Answer D is incorrect. Both Technicians are correct.

TASK A.8

35. The LEAST LIKELY cause of low cylinder compression is:

    A. Worn valves
    B. Worn rings
    C. Blown head gasket
    D. Worn valve guides

    Answer A is incorrect. Worn valves will not properly seal the combustion chamber and will cause a compression loss.

    Answer B is incorrect. Worn piston rings are a common cause of low cylinder compression.

    Answer C is incorrect. A blown head gasket can cause a loss of cylinder compression.

    Answer D is correct. Worn valve guides often cause excessive oil consumption but may not affect valve sealing; therefore, they rarely cause lower compression.

TASK H.9

36. Technician A says that during a PCM replacement, the technician should use a grounding strap to prevent static charge from damaging the PCM. Technician B says care should be taken not to touch the PCM terminals with your fingers. Who is correct?

    A. A only
    B. B only
    C. Both A and B
    D. Neither A nor B

    Answer A is incorrect. Technician B is also correct.

    Answer B is incorrect. Technician A is also correct.

    Answer C is correct. Both Technicians are correct. Any static discharge of electricity can damage the PCM, and using a ground strap helps prevent this from happening. You should never touch the terminals of a PCM.

    Answer D is incorrect. Both Technicians are correct.

TASK F.3

37. Technician A says that a nylon fuel line that is bent sharply, causing a kink, is allowable and will not affect fuel flow. Technician B says nylon fuel line cannot be repaired, and the entire line must be replaced. Who is correct?

    A. A only
    B. B only
    C. Both A and B
    D. Neither A nor B

    Answer A is incorrect. A kink in a fuel line can cause reduced fuel flow and aeration that may cause drivability problems.

    Answer B is incorrect. Nylon fuel line can be spliced and end connectors replaced using approved repair kits.

    Answer C is incorrect. Neither Technician is correct.

    Answer D is correct. Neither Technician is correct. A kink in a fuel line can cause reduced fuel flow and aeration that may cause drivability problems. Nylon fuel line can be spliced and end connectors replaced using approved repair kits.

38. During a cylinder power balance test, there is no RPM drop on cylinder #3. Technician A says that the cylinder may have an inoperative injector. Technician B says that the cylinder may have an inoperative spark plug. Who is correct?

TASK A.7

   A. A only
   B. B only
   C. Both A and B
   D. Neither A nor B

Answer A is incorrect. Technician B is also correct.

Answer B is incorrect. Technician A is also correct.

Answer C is correct. Both Technicians are correct. Any cylinder that does not produce an RPM drop relative to the other cylinders during a power balance test indicates the cylinder is not producing equal power and has a problem with the ignition, fuel delivery, or mechanical integrity of the cylinder. An inoperative spark plug or inoperative injector could cause no RPM drop during a power balance test.

Answer D is incorrect. Both Technicians are correct.

39. The LEAST LIKELY symptom resulting from an evaporative emissions (EVAP) system failure is:

TASK H.4

   A. Increased tailpipe emissions
   B. Low vehicle emissions
   C. Malfunction indicator lamp (MIL) illumination
   D. Fuel odor

Answer A is incorrect. EVAP system failure can cause purging of the canister during idle or a saturated charcoal canister, which can increase tailpipe emissions.

Answer B is correct. A malfunction would usually result in increased emissions.

Answer C is incorrect. EVAP system problems may set DTCs and illuminate the MIL.

Answer D is incorrect. Fuel odor is a common indication of EVAP system leaks.

40. A pulse-style, secondary air-injection system is driven:

TASK G.6

   A. By a drive belt
   B. Hydraulically
   C. By an electric motor
   D. With negative pressure pulses in the exhaust system

Answer A is incorrect. Some secondary air-injection systems use a belt-driven pump. Belt-style secondary air-injection systems use a belt, not pulse air.

Answer B is incorrect. There are currently no hydraulically-operated air-injection systems.

Answer C is incorrect. Some secondary air-injection systems use a pump driven by an electric motor.

Answer D is correct. A pulse air-injection system works because of the negative pressure pulses in the exhaust system.

TASK F.10

41. A truck that has low power also has low intake manifold vacuum. Technician A says a restricted air filter could be the cause. Technician B says a restricted exhaust could be the cause. Who is correct?

    A. A only

    B. B only

    C. Both A and B

    D. Neither A nor B

    Answer A is incorrect. A restricted air filter would increase the vacuum in the intake manifold.

    Answer B is correct. Only Technician B is correct. A restricted exhaust system would slow airflow in the exhaust, which in turn would slow the airflow through the intake manifold and cause the intake manifold to be reduced and the engine to have a low power concern.

    Answer C is incorrect. Technician A is incorrect.

    Answer D is incorrect. Technician B is correct.

TASK E.5

42. Technician A says a thorough ignition coil test includes both primary and secondary winding resistance tests. Technician B says maximum coil output testing can be performed with an oscilloscope. Who is correct?

    A. A only

    B. B only

    C. Both A and B

    D. Neither A nor B

    Answer A is incorrect. Technician B is also correct.

    Answer B is incorrect. Technician A is also correct.

    Answer C is correct. Both Technicians are correct. The coil has two circuits, a primary and secondary. Both need to be tested for resistance. A dynamic output test using an oscilloscope will show specific voltages.

    Answer D is incorrect. Both Technicians are correct.

TASK G.12

43. Technician A says the filter on the engine off, natural vacuum (EONV) EVAP system should be serviced at regular oil change intervals. Technician B says the EONV EVAP system does not use a vacuum pump. Who is correct?

    A. A only

    B. B only

    C. Both A and B

    D. Neither A nor B

    Answer A is incorrect. The EONV EVAP system does use a filter. However, it is not changed as often as the engine oil.

    Answer B is correct. Only Technician B is correct. The EONV system does not use a pump.

    Answer C is incorrect. Only Technician B is correct.

    Answer D is incorrect. Technician B is correct.

44. The leak down detection system is designed to:

    A. Pressurize the EVAP system to check for leaks
    B. Control the idle speed
    C. Transfer fuel from the tank to the injectors
    D. Stabilize engine RPM under a load

    Answer A is correct. A leak down detection system is designed to pressurize the EVAP system to check for leaks.

    Answer B is incorrect. Idle speed control is performed with the idle air control or the computer-controlled throttle body, not the leak down detection system.

    Answer C is incorrect. A leak down detection system does not transfer fuel; that is the job of the fuel supply system.

    Answer D is incorrect. The computer, not the leak down detection system, controls engine RPM by controlling fuel, spark and in some cases air.

TASK G.10

45. During a cylinder leak down test on a 4-cylinder engine, air is heard coming from spark plug hole #3 as cylinder #4 is being checked. Technician A says that this could be caused by a blown head gasket. Technician B says this could be caused by a cracked engine block. Who is correct?

    A. A only
    B. B only
    C. Both A and B
    D. Neither A nor B

    Answer A is incorrect. Technician B is also correct.

    Answer B is incorrect. Technician A is also correct.

    Answer C is correct. Both Technicians are correct. If there is a leak in an adjacent cylinder, the technician should look at what is "common" to both cylinders. A leaking head gasket or a cracked engine block between cylinder #3 and #4 would allow air to pass from cylinder #3 to cylinder #4 during a cylinder leak down test.

    Answer D is incorrect. Both Technicians are correct.

TASK A.9

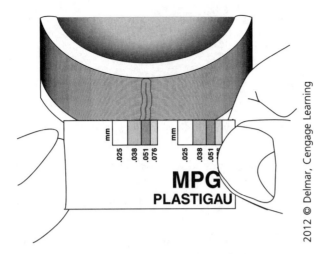

2012 © Delmar, Cengage Learning

TASK C.7

46. In the above illustration, which of the following would be true?

    A. Crank end-play is 0.051 mm.

    B. Bearing crush is 0.051 mm.

    C. Bearing clearance is 0.051 mm.

    D. Crank runout is 0.051 mm.

    Answer A is incorrect. Crank end-play is not being measured. A dial indicator is used to measure crank end-play.

    Answer B is incorrect. Bearing crush is not measured with Plastigauge. Bearing crush is not measured during engine rebuild.

    Answer C is correct. Plastigauge is commonly used to check friction bearing clearance in the manner shown in the figure, and the clearance measures 0.051 mm.

    Answer D is incorrect. Crank runout is not being measured. Crank runout is measured with a dial indicator.

TASK A.4

47. An engine is making a loud metallic knocking that gets louder as the engine warms up or if the throttle is quickly snapped open. The noise almost disappears when the spark for cylinder #3 is shorted to ground. Technician A says the symptoms indicate a cracked flywheel. Technician B says the symptoms indicate a loose connecting-rod bearing. Who is correct?

    A. A only

    B. B only

    C. Both A and B

    D. Neither A nor B

    Answer A is incorrect. A cracked flywheel would not be sensitive to engine temperature and would not diminish when shorting a spark plug wire.

    Answer B is correct. Only Technician B is correct. A loose connecting-rod bearing will produce a sharp metallic noise that gets louder as the engine warms up and the oil thins out or when cylinder pressure increases from a throttle snap. Shorting out the cylinder's spark plug often makes the noise go away or become much quieter.

    Answer C is incorrect. Only Technician B is correct.

    Answer D is incorrect. Technician B is correct.

48. A vehicle with an electronic ignition fails to start. Technician A says this could be caused by a defective crankshaft sensor connection. Technician B says this could be caused by a defective ignition module. Who is correct?

TASK F.1, E.3

   A.   A only

   B.   B only

   C.   Both A and B

   D.   Neither A nor B

Answer A is incorrect. Technician B is also correct.

Answer B is incorrect. Technician A is also correct.

Answer C is correct. Both Technicians are correct. Any interruption of the primary input signals will not switch the module. A defective crankshaft sensor connection can prevent the ignition module from receiving the signal. A defective ignition module can result in no spark at the spark plug.

Answer D is incorrect. Both Technicians are correct.

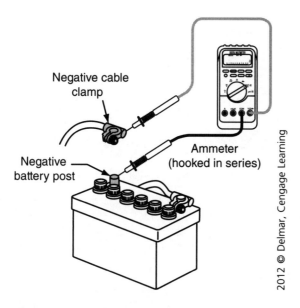

Negative cable clamp

Ammeter (hooked in series)

Negative battery post

2012 © Delmar, Cengage Learning

49. An ammeter, set in the milliamp position, is connected in series between the negative battery cable and ground, as shown in the above illustration. What is being measured?

TASK A.14

   A.   Starter draw

   B.   Battery drain

   C.   Regulated voltage

   D.   Voltage drop

Answer A is incorrect. Starter draw is checked in amps.

Answer B is correct. Any "live" circuit will cause amperage flow.

Answer C is incorrect. Voltage is measured in volts.

Answer D is incorrect. Voltage drops are measured in volts, not milliamps.

TASK F.1

50. A vehicle with a no-start condition is being diagnosed. The vehicle has 12 volts at the fuel pump connector while cranking, but the fuel pump does not run. Technician A says the fuel pump could be the cause. Technician B says the ground side of the fuel pump circuit should be checked for open or high resistance. Who is correct?

    A. A only
    B. B only
    C. Both A and B
    D. Neither A nor B

    Answer A is incorrect. Technician B is also correct.

    Answer B is incorrect. Technician A is also correct.

    Answer C is correct. Both Technicians are correct. The fuel pump could be defective, but it should not be replaced without checking the positive and ground side of the fuel pump for voltage drop or an open circuit.

    Answer D is incorrect. Both Technicians are correct.

PREPARATION EXAM 6—ANSWER KEY

1.	C	21.	C	41.	A
2.	C	22.	D	42.	A
3.	D	23.	C	43.	C
4.	A	24.	D	44.	A
5.	D	25.	A	45.	A
6.	C	26.	C	46.	C
7.	D	27.	C	47.	B
8.	C	28.	D	48.	A
9.	B	29.	C	49.	D
10.	A	30.	B	50.	C
11.	B	31.	D		
12.	C	32.	D		
13.	C	33.	C		
14.	A	34.	D		
15.	C	35.	A		
16.	B	36.	C		
17.	B	37.	C		
18.	C	38.	B		
19.	D	39.	D		
20.	B	40.	C		

PREPARATION EXAM 6—EXPLANATIONS

1. A technician is performing a running compression test on a vehicle with suspected cylinder sealing problems. Technician A says running compression should be half of static compression at idle. Technician B says during a running compression test, the technician should snap the throttle, and the running compression should be 80 percent of static compression. Who is correct?

TASK A.8

  A. A only
  B. B only
  C. Both A and B
  D. Neither A nor B

  Answer A is incorrect. Technician B is also correct.

  Answer B is incorrect. Technician A is also correct.

  Answer C is correct. Both Technicians are correct. Due to the relatively closed throttle plate, running compression should be half of static compression when tested at idle and 80 percent of static compression when the throttle is snapped.

  Answer D is incorrect. Both Technicians are correct.

TASK E.4

2. Technician A says the secondary ignition system is designed to handle voltages as high as 90,000 volts. Technician B says low secondary system voltage output could be the result of high primary ignition circuit resistance. Who is correct?

   A. A only
   B. B only
   C. Both A and B
   D. Neither A nor B

   Answer A is incorrect. Technician B is also correct.

   Answer B is incorrect. Technician A is also correct.

   Answer C is correct. Both Technicians are correct. The secondary ignition is designed to handle as much as 90,000 volts or more. High primary circuit resistance will reduce primary circuit current flow, which in turn would reduce the secondary output voltage.

   Answer D is incorrect. Both Technicians are correct.

TASK C.11

3. Which of the following steps is the technician LEAST LIKELY to perform when pressing the wrist pin into the piston and connecting rods?

   A. Align the bores in the piston and connecting rod.
   B. Heat the small end of the rod.
   C. Make sure that position marks on the piston and connecting rod are oriented properly.
   D. Heat the wrist pin.

   Answer A is incorrect. The bores in the piston and the rod should be aligned before pressing in the wrist pin.

   Answer B is incorrect. The small end of the rod should be heated before pressing in the wrist pin.

   Answer C is incorrect. Position marks on the piston and rod should be oriented properly before pressing in the wrist pin.

   Answer D is correct. Heating the wrist pin is a step LEAST LIKELY performed by the technician. Heating the wrist pin will cause it to expand and prevent it from fitting into the connecting rod and piston bores.

TASK F.10

4. Technician A says some vehicles require that the minimum airflow be adjusted at a throttle body service. Technician B says that some vehicles require that the minimum airflow be adjusted at the manifold absolute pressure sensor. Who is correct?

   A. A only
   B. B only
   C. Both A and B
   D. Neither A nor B

   Answer A is correct. Only Technician A is correct. Some vehicles will have a minimum airflow adjustment on the throttle body. This adjustment sets the minimum throttle opening.

   Answer B is incorrect. The manifold absolute pressure will not have a minimum airflow adjustment. The sensor simply measures the amount of intake vacuum.

   Answer C is incorrect. Only Technician A is correct.

   Answer D is incorrect. Technician A is correct.

5. A truck has an occasional stumble on acceleration. There are no diagnostic trouble codes (DTCs) set. Which of the following could be the cause?

    A. Restricted exhaust

    B. Restricted air filter

    C. Open coolant temperature sensor

    D. Faulty APP sensor

TASK H.5

Answer A is incorrect. Restricted exhaust will cause low power, not an occasional stumble on acceleration.

Answer B is incorrect. A restricted air filter will cause low power, not an occasional stumble on acceleration.

Answer C is incorrect. An open coolant temperature sensor would set a DTC.

Answer D is correct. A worn APP may not set a DTC if it is an in-range failure, which may cause the PCM to misread the pedal position and cause a stumble.

6. Technician A says the exhaust gas recirculation (EGR) system is used to lower combustion chamber temperature. Technician B says that EGR systems that use an EGR valve position sensor should read about 4.5 volts with the EGR valve at full open. Who is correct?

    A. A only

    B. B only

    C. Both A and B

    D. Neither A nor B

TASK G.3

Answer A is incorrect. Technician B is also correct.

Answer B is incorrect. Technician A is also correct.

Answer C is correct. Both Technicians are correct. The EGR system is used to control oxides of nitrogen (NOx). It does this by lowering the combustion chamber temperature. If the EGR system uses an EGR position sensor, it would read about 4.5 volts with the valve at full open and about 1 volt with the valve fully closed.

Answer D is incorrect. Both Technicians are correct.

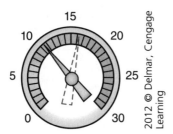

TASK A.6

7. An engine is idling at 750 rpm. The pointer on the vacuum gauge in the above illustration is floating between 11 and 16 in. Hg. The most likely cause would be:

A. Retarded timing
B. Advanced timing
C. A stuck EGR system valve
D. Too lean an idle mixture

Answer A is incorrect. Retarded timing would not result in gauge fluctuation; it would result in a low steady gauge.

Answer B is incorrect. Advanced timing would not result in gauge fluctuation; it would result in a higher than normal reading.

Answer C is incorrect. A stuck EGR valve would not result in gauge fluctuation; it would result in a pulsing needle or a low reading, depending on the location of the EGR valve.

Answer D is correct. A lean air/fuel mixture can result in a pulsing needle. The needle pulses because of the fluctuation in vacuum in the intake manifold.

TASK H.4

8. An engine equipped with electronic fuel injection has a loose exhaust manifold. Technician A says that the loose manifold may cause poor drivability. Technician B says that the loose manifold may cause an oxygen sensor code to be set. Who is correct?

A. A only
B. B only
C. Both A and B
D. Neither A nor B

Answer A is incorrect. Technician B is also correct.

Answer B is incorrect. Technician A is also correct.

Answer C is correct. Both Technicians are correct. On an engine equipped with electronic fuel injection, a loose exhaust manifold may cause the engine to stumble, stall, and backfire due to the fresh air entering the exhaust stream. It may also cause the oxygen sensor to set a lean air/fuel mixture code, as a loose manifold will allow fresh air into the exhaust stream. This condition will lead to poor vehicle drivability.

Answer D is incorrect. Both Technicians are correct.

TASK B.2

9. A vehicle with DTC PO302 cylinder #2 misfire is being diagnosed. Technician A says low fuel pump pressure could be the cause. Technician B says a fouled spark plug could be the cause. Who is correct?

A. A only
B. B only
C. Both A and B
D. Neither A nor B

Answer A is incorrect. Low fuel pressure would affect all cylinders, not just one.

Answer B is correct. Only Technician B is correct. A fouled spark plug could cause a single cylinder misfire code.

Answer C is incorrect. Only Technician B is correct.

Answer D is incorrect. Technician B is correct.

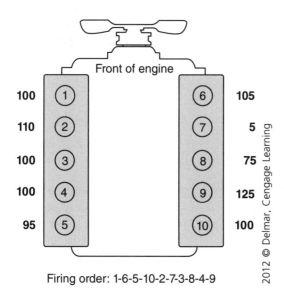

Front of engine

100 ①   ⑥ 105

110 ②   ⑦ 5

100 ③   ⑧ 75

100 ④   ⑨ 125

95 ⑤   ⑩ 100

2012 © Delmar, Cengage Learning

Firing order: 1-6-5-10-2-7-3-8-4-9

10. Cylinder power balance test results from a fuel-injected engine with a coil-on-plug ignition are shown in the above illustration. One cylinder is found to have virtually no RPM change. Which of these is the most likely cause?

TASK E.1

    A. A faulty fuel injector

    B. A faulty crank position sensor

    C. A vacuum leak at the throttle body

    D. A fuel saturated vapor canister

Answer A is correct. A faulty injector could fail to deliver fuel and result in no power from a single cylinder.

Answer B is incorrect. A faulty crank sensor would not cause only one cylinder to not contribute in a power balance test.

Answer C is incorrect. A leak here would affect fuel mixture on all cylinders.

Answer D is incorrect. The fuel from the vapor canister would affect all cylinders.

11. A vehicle is being diagnosed for a rough idle. The vehicle exhaust has no visible color; however, the exhaust has a very strong odor and burns the eyes. Technician A says the vehicle could have a stuck closed pressure regulator, causing an excessively lean mixture. Technician B says the vehicle could have a restricted fuel filter, causing it to run too lean. Who is correct?

TASK A.5

    A. A only

    B. B only

    C. Both A and B

    D. Neither A nor B

Answer A is incorrect. A stuck closed fuel-pressure regulator would cause a rich exhaust. A rich exhaust would cause black smoke.

Answer B is correct. Only Technician B is correct. Because of the colorless exhaust and the burning-the-eyes effect, the vehicle is probably running lean. A restricted fuel filter, vacuum leak, or weak fuel pump could cause these problems.

Answer C is incorrect. Only Technician B is correct.

Answer D is incorrect. Technician B is correct.

TASK C.15

12. Technician A says a special puller and installer tool may be required to remove and install the vibration damper. Technician B says if the inertia ring on the vibration damper is loose, the damper must be replaced. Who is correct?

A. A only

B. B only

C. Both A and B

D. Neither A nor B

Answer A is incorrect. Technician B is also correct.

Answer B is incorrect. Technician A is also correct.

Answer C is correct. Both Technicians are correct. A special puller and installer tool are required to remove and install the vibration damper. Using a regular gear puller to remove the vibration damper will damage the damper. A loose inertia ring on the damper requires replacement of the damper.

Answer D is incorrect. Both Technicians are correct.

TASK H.5

13. A truck has a short-term trim value that indicates the PCM is adding fuel. There is also a strong sulfur smell in the vehicle's exhaust. Which of the following is the most likely cause?

A. A lean fuel mixture

B. Coolant leaking into a combustion chamber

C. A rich fuel mixture

D. A vacuum leak

Answer A is incorrect. A lean fuel mixture could cause a misfire, but it would not cause a sulfur smell.

Answer B is incorrect. Coolant leaking into the combustion chambers would cause a gray exhaust color.

Answer C is correct. A sulfur smell is commonly due to raw fuel in the converter from an overly rich fuel mixture.

Answer D is incorrect. A vacuum leak would cause a rough idle that would decrease as engine speed increases.

TASK A.8

14. An engine that has a single cylinder misfire is being diagnosed by performing a leak down test. Which of these is the most likely cause of the condition?

A. 45 percent leakage with air coming out of the intake valve

B. 10 percent leakage with air coming out of the crankcase

C. 5 percent leakage with air coming out of the crankcase

D. 10 percent leakage with air coming out of the exhaust valve

Answer A is correct. 45 percent leakage is excessive and could cause be the cause of a single cylinder misfire.

Answer B is incorrect. 10 percent leakage is acceptable and would not cause a misfire.

Answer C is incorrect. 5 percent leakage is acceptable and would not cause a misfire.

Answer D is incorrect. 10 percent leakage is acceptable and would not cause a misfire.

15. Technician A says that secondary air is injected upstream of the converter on a cold engine to aid in heating up the oxygen sensor. Technician B says secondary air is injected downstream to the converter on a warm engine to aid in catalytic converter operation. Who is correct?

TASK G.8

   A. A only
   B. B only
   C. Both A and B
   D. Neither A nor B

   Answer A is incorrect. Technician B is also correct.

   Answer B is incorrect. Technician A is also correct.

   Answer C is correct. Both Technicians are correct. Secondary air is injected upstream on a cold engine and downstream on a warm engine. The airflow is switched to optimize emission control.

   Answer D is incorrect. Both Technicians are correct.

16. A Hall effect sensor is being tested. Technician A says the Hall effect sensor should have a resistance value of 500 to 800 ohms. Technician B says the use of a lab scope is an accurate method of checking Hall effect sensor operation. Who is correct?

TASK E.3

   A. A only
   B. B only
   C. Both A and B
   D. Neither A nor B

   Answer A is incorrect. Hall effect sensors do not have a resistance value and cannot be measured for resistance.

   Answer B is correct. Only Technician B is correct. A lab scope is the best way to tell if a Hall effect sensor is producing a signal.

   Answer C is incorrect. Only Technician B is correct.

   Answer D is incorrect. Technician B is correct.

17. There is a slight vapor from the exhaust accompanied by a sweet odor. Which of the following could be the cause?

TASK A.5

   A. A leaking exhaust manifold
   B. A leaking intake manifold
   C. A faulty barometric pressure (BARO) sensor
   D. A faulty accelerator pedal position (APP) sensor

   Answer A is incorrect. A leaking exhaust manifold would not cause a sweet odor or a vapor in the exhaust stream. A leaking exhaust manifold can cause a noise and the smell of exhaust in the truck cab.

   Answer B is correct. A leaking intake manifold can allow coolant into the combustion chamber, which can cause a sweet odor and a slight vapor.

   Answer C is incorrect. A faulty BARO sensor can cause low power but will not cause a sweet-smelling odor.

   Answer D is incorrect. A faulty APP sensor can cause erratic throttle operation but would not cause a sweet odor in the exhaust.

TASK F.12

18. A truck with an under-vehicle rattle and low engine power is being diagnosed. Technician A says an intake manifold vacuum test could be performed to help locate a restricted exhaust. Technician B says the catalytic converter may have come apart and may be restricting exhaust flow. Who is correct?

A. A only

B. B only

C. Both A and B

D. Neither A nor B

Answer A is incorrect. Technician B is also correct.

Answer B is incorrect. Technician A is also correct.

Answer C is correct. Both Technicians are correct. If an engine has low vacuum at a sustained engine RPM of 2500, a restricted exhaust can be the cause. The exhaust system rattle could be the result of loose components in the catalytic converter causing a restricted exhaust flow. A restriction in the exhaust can cause a loss of power on acceleration and cruise.

Answer D is incorrect. Both Technicians are correct.

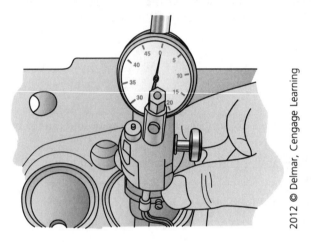

2012 © Delmar, Cengage Learning

TASK B.4

19. In the illustration above, the technician is most likely checking:

A. Valve guide depth

B. Valve seat angle

C. Cylinder head flatness

D. Valve seat runout

Answer A is incorrect. Valve guide depth is measured with a depth micrometer.

Answer B is incorrect. The tool shown does not measure valve seat angle. Valve seat angle can be checked by lightly touching the seat with an appropriately angled stone and checking the contact pattern.

Answer C is incorrect. A straightedge and a feeler gauge are used to measure the cylinder head.

Answer D is correct. This is a valve seat concentricity gauge. It is used to check valve seat runout.

20. A vehicle is being diagnosed for an overheating complaint while at cruising speeds. Technician A says the electric cooling fan may be defective. Technician B says that the radiator may have some damaged cooling fins. Who is correct?

    A. A only
    B. B only
    C. Both A and B
    D. Neither A nor B

TASK D.6

Answer A is incorrect. The fan is used to circulate air through the radiator when sitting still or moving slowly in traffic. A faulty fan would cause overheating in traffic or when sitting still, but not when cruising.

Answer B is correct. Only Technician B is correct. A radiator with damaged cooling fins could cause an overheating problem while cruising.

Answer C is incorrect. Only Technician B is correct.

Answer D is incorrect. Technician B is correct.

21. Oil is leaking from an engine's crankshaft rear main bearing seal. Technician A says the oil seal could be faulty. Technician B says the positive crankcase ventilation (PCV) system may not be functioning. Who is correct?

    A. A only
    B. B only
    C. Both A and B
    D. Neither A nor B

TASK A.3

Answer A is incorrect. Technician B is also correct.

Answer B is incorrect. Technician A is also correct.

Answer C is correct. Both Technicians are correct. The cause of an oil leak from the crankshaft rear main bearing seal could be a faulty oil seal or a malfunctioning PCV system. A faulty system will cause incorrect crankcase pressures.

Answer D is incorrect. Both Technicians are correct.

22. If new rings are installed without removing the ring ridge, which of these is the most likely result?

    A. Piston skirt damage
    B. Piston pin damage
    C. Connecting-rod bearing damage
    D. Piston compression ring damage

TASK C.3

Answer A is incorrect. The piston skirt would not be damaged.

Answer B is incorrect. The piston pin is not affected by the ring ridge.

Answer C is incorrect. The connecting-rod bearings would be unaffected.

Answer D is correct. Failure to remove the ring ridge may cause piston ring and/or top compression ring damage after the engine is assembled and started.

23. Technician A says most 4-wire COP (coil-on-plug) ignition systems use individual ignition control modules for each cylinder. Technician B says a two-wire COP coil is tested like any other ignition coil. Who is correct?

    A. A only
    B. B only
    C. Both A and B
    D. Neither A nor B

    Answer A is incorrect. Technician B is also correct.

    Answer B is incorrect. Technician A is also correct.

    Answer C is correct. Both Technicians are correct. A four-wire coil-on-plug coil will have the ignition control module integrated with it. A two-wire COP will not have an integrated module, and the coil can be checked like a standard ignition coil.

    Answer D is incorrect. Both Technicians are correct.

24. The customer is concerned by a loud popping noise, most noticeable during acceleration. Which of these is the most likely cause?

    A. An intake manifold leak
    B. A cooling system leak
    C. A missing air filter
    D. An exhaust manifold leak

    Answer A is incorrect. An intake manifold leak, if it can be heard, would be more of a rushing-air or whistling noise.

    Answer B is incorrect. A cooling system leak usually does not make a sound, but it will usually make a wet spot.

    Answer C is incorrect. A missing air filter would not cause a popping noise upon acceleration, but it can cause drivability problems related to the mass airflow sensor.

    Answer D is correct. An exhaust manifold leak has a definite popping sound that is much more noticeable upon acceleration.

25. Technician A says the powertrain control module (PCM) controls the amount of canister purge. Technician B says the evaporative emissions (EVAP) system controls the amount of NO_X emissions produced by the engine. Who is correct?

    A. A only
    B. B only
    C. Both A and B
    D. Neither A nor B

    Answer A is correct. Only Technician A is correct. The PCM controls canister purge.

    Answer B is incorrect. The EGR system controls the amount of NO_X.

    Answer C is incorrect. Only Technician A is correct.

    Answer D is incorrect. Technician A is correct.

26. The maximum secondary voltage from an ignition coil is lower than specification. Which of the following could be the cause?

    A. An open primary winding
    B. An open secondary winding
    C. High resistance in the primary winding
    D. High resistance in the secondary ignition circuit.

    TASK E.3

    Answer A is incorrect. An open primary winding would result in zero secondary voltage output.

    Answer B is incorrect. An open secondary winding would result in zero secondary voltage output.

    Answer C is correct. High resistance in the primary winding would cause a weak field to be generated in the primary windings and therefore a low secondary voltage output.

    Answer D is incorrect. High resistance in the secondary ignition circuit would cause an increase in secondary voltage output because a higher voltage would be required to jump the spark plug gap.

27. A vehicle with a rich exhaust code is being diagnosed. Technician A says the fuel pressure could be the cause. Technician B says a ruptured fuel-pressure regulator diaphragm could be the cause. Who is correct?

    A. A only
    B. B only
    C. Both A and B
    D. Neither A nor B

    TASK F.6

    Answer A is incorrect. Technician B is also correct.

    Answer B is incorrect. Technician A is also correct.

    Answer C is correct. Both Technicians are correct. High fuel pressure or a ruptured diaphragm in a fuel-pressure regulator could both cause a rich exhaust code.

    Answer D is incorrect. Both Technicians are correct.

28. After a vehicle is parked overnight and then started in the morning, the engine has a lifter noise that disappears after it has been running for a short while. The most likely cause would be:

    A. Low oil pressure
    B. Low oil level
    C. A worn lifter bottom
    D. Excessive lifter leak down

    TASK A.4

    Answer A is incorrect. Low oil pressure would result in a continuous noise.

    Answer B is incorrect. Low oil level would result in a continuous noise.

    Answer C is incorrect. A worn lifter bottom would result in continuous noise.

    Answer D is correct. Excessive lifter leak down, usually caused by a dirty or leaking check ball in the lifter, can cause this concern.

29. Technician A says the PCV valve flow volume is low at wide-open throttle (WOT). Technician B says the PCV valve flow volume is low during deceleration. Who is correct?

    A. A only
    B. B only
    C. Both A and B
    D. Neither A nor B

    TASK G.2

    Answer A is incorrect. Technician B is also correct.

    Answer B is incorrect. Technician A is also correct.

    Answer C is correct. Both Technicians are correct. The PCV valves are restricted at idle, WOT, and during deceleration. The flow is higher at times of lower intake manifold vacuum, such as at cruise speeds.

    Answer D is incorrect. Both Technicians are correct.

TASK B.4

30. Technician A says valve stem height is measured from the bottom of the valve guide to the top of the valve guide. Technician B says that excessive valve stem-to-guide clearance may result in excessive oil consumption. Who is correct?

    A.   A only
    B.   B only
    C.   Both A and B
    D.   Neither A nor B

Answer A is incorrect. Valve stem height is usually measured between the spring seat and the tip of the valve.

Answer B is correct. Only Technician B is correct. If valve stem-to-guide clearance is excessive, the vacuum created by the downward movement of the piston may draw oil that is around the guide through the guide.

Answer C is incorrect. Only Technician B is correct.

Answer D is incorrect. Technician B is correct.

2012 © Delmar, Cengage Learning

TASK A.8

31. When using a compression tester, as shown in the above illustration, the compression readings on the cylinders are all even, but lower than the specified compression. This could indicate:

    A.   A blown head gasket
    B.   Carbon buildup
    C.   A cracked head
    D.   Worn rings and cylinders

Answer A is incorrect. A low reading on two adjacent cylinders may indicate a blown head gasket.

Answer B is incorrect. Carbon buildup would cause a high reading.

Answer C is incorrect. A low reading on two adjacent cylinders may indicate a cracked cylinder head.

Answer D is correct. Worn rings and cylinders can result in low compression on all cylinders.

TASK E.3

32. On a truck equipped with coil-on-plug ignition, one primary circuit current flow is higher than specification. Which of the following could be the cause?

    A.   High primary coil resistance
    B.   High secondary coil resistance
    C.   Low secondary coil resistance
    D.   Low primary coil resistance

Answer A is incorrect. High primary coil resistance would cause low current flow when resistance increase current decreases.

Answer B is incorrect. Secondary coil resistance does not affect primary coil current flow.

Answer C is incorrect. Secondary coil resistance does not affect primary coil current flow.

Answer D is correct. Low primary coil resistance can cause an increase in current flow.

33. A powertrain control module is being replaced. Technician A says the new PCM should be ordered using the original PCM's part number. Technician B says installing a used PCM may cause the theft deterrent system to activate on some vehicles. Who is correct?

TASK H.9

   A. A only
   B. B only
   C. Both A and B
   D. Neither A nor B

Answer A is incorrect. Technician B is also correct.

Answer B is incorrect. Technician A is also correct.

Answer C is correct. Both Technicians are correct. Due to emission calibration and engine variables, the same PCM part number or the newest superseded number must be used. On some vehicles, if a used PCM is installed, it causes the theft deterrent system to activate, causing a no-start.

Answer D is incorrect. Both Technicians are correct.

34. A cylinder power balance test is being performed on an engine to determine which cylinder is causing a misfire. Technician A says that when the faulty cylinder is disabled, engine RPM will drop more than for the other cylinders. Technician B says that disabling the faulty cylinder will cause the engine to stall. Who is correct?

TASK A.7

   A. A only
   B. B only
   C. Both A and B
   D. Neither A nor B

Answer A is incorrect. Disabling the faulty cylinder will cause engine RPM to drop less than for the other cylinders.

Answer B is incorrect. Disabling the faulty cylinder will not cause the engine to stall.

Answer C is incorrect. Neither Technician is correct.

Answer D is correct. Neither Technician is correct. Disabling the faulty cylinder will cause engine RPM to drop less that for the other cylinders, but it will not cause the engine to stall.

35. The customer says that the engine requires excessive cranking to start. The LEAST LIKELY cause of this problem would be:

TASK F.1

   A. A cracked cylinder block
   B. A jumped timing belt
   C. A faulty fuel pump
   D. A stuck open EGR valve

Answer A is correct. The least likely cause is a cracked block. A cracked block can cause coolant/oil mixing and/or external leaks, which would not affect cranking.

Answer B is incorrect. If valve timing is off, long cranks or a no-start condition may be the result.

Answer C is incorrect. A faulty pump could cause long crank time due to low fuel flow and pressure.

Answer D is incorrect. A stuck open EGR valve could cause long crank times due to the lack of vacuum.

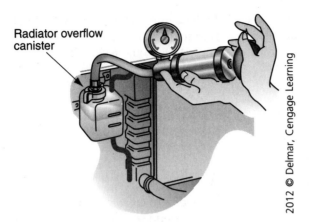

Radiator overflow canister

2012 © Delmar, Cengage Learning

TASK D.6

36. The tester in the illustration above may be used to test all the following items EXCEPT:

A.   Coolant leaks

B.   The radiator cap pressure relief valve

C.   Freeze protection level

D.   Heater core leaks

Answer A is incorrect. The tester may be used to test for cooling system leaks.

Answer B is incorrect. The tester may be used to test the radiator cap pressure release valve.

Answer C is correct. This tester is not used to test for freeze protection level. A refractometer or ball float gauge is used to check coolant freeze protection.

Answer D is incorrect. The tester may be used to test for heater core leaks.

TASK H.6

37. A throttle position sensor (TPS) is being tested with a digital multi-meter (DMM). Technician A says the voltage on the signal wire should be around 1.0 volt or less at idle. Technician B says the voltage reading should be watched for irregularities while sweeping the TPS. Who is correct?

A.   A only

B.   B only

C.   Both A and B

D.   Neither A nor B

Answer A is incorrect. Technician B is also correct.

Answer B is incorrect. Technician A is also correct.

Answer C is correct. Both Technicians are correct. The TPS should read about 1.0 volt or less at idle, and it should be sweep tested.

Answer D is incorrect. Both Technicians are correct.

TASK H.7

38. A voltage drop test is being performed on an ignition coil power supply wire. When performing this test:

A.   The key should be off.

B.   The engine should be running.

C.   The key should be on, but the engine should not be running.

D.   The vehicle should be driven above 35 mph.

Answer A is incorrect. To perform a voltage drop test, the circuit must be energized and operating under load.

Answer B is correct. This will place a full load on the power supply circuit and will cause a voltage drop to appear if a problem with the circuit is present.

Answer C is incorrect. This will energize the circuit; however, the load will not be as high as it will be when the engine is running.

Answer D is incorrect. There is no reason to drive the vehicle during this test, as a sufficient load will be placed on the circuit by operating the engine with the vehicle parked.

39. A vehicle is being diagnosed for a PO134 DTC (oxygen sensor circuit, no activity detected on bank 1, sensor 1). Technician A says the problem could be low fuel system pressure. Technician B says the problem could be the downstream oxygen sensor on the bank where the #1 cylinder is located. Who is correct?

TASK H.2

   A. A only
   B. B only
   C. Both A and B
   D. Neither A nor B

Answer A is incorrect. Low fuel pressure would affect all sensor readings, not just one.

Answer B is incorrect. Bank 1, sensor 1 is the upstream sensor on the bank where the #1 cylinder is located.

Answer C is incorrect. Neither Technician is correct.

Answer D is correct. Neither Technician is correct. Low fuel pressure would affect all sensor readings, not just one. Bank 1, sensor 1 is the upstream sensor on the bank where cylinder #1 is located. This code is most likely caused by a failed upstream oxygen sensor on bank 1.

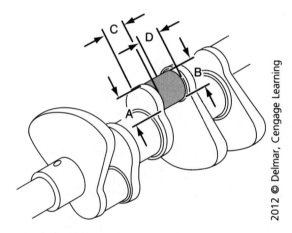

2012 © Delmar, Cengage Learning

40. When measuring the crankshaft journal, as shown in the above illustration, the difference between measurements:

TASK C.6

   A. A and B indicates out of round.
   B. C and D indicates out of round.
   C. A and C indicates out of round.
   D. A and D indicates taper.

Answer A is incorrect. Measurements A and B indicate taper.

Answer B is incorrect. Measurements C and D indicate taper.

Answer C is correct. The difference in measurement between A and C indicates out of round.

Answer D is incorrect. Measurements A and D give an indication of general crankshaft condition, but they do not indicate taper.

TASK F.8

41. Technician A says group-fired injection systems fire more than one injector at the same time. Technician B says sequential fuel injection systems fire two or more injectors at the same time. Who is correct?

    A. A only
    B. B only
    C. Both A and B
    D. Neither A nor B

    Answer A is correct. Only Technician A is correct. In a group-fired injection system, more than one injector is fired at the same time.

    Answer B is incorrect. Sequential fuel injection systems fire only one injector at a time.

    Answer C is incorrect. Only Technician A is correct.

    Answer D is incorrect. Technician A is correct.

TASK B.8

42. On an overhead camshaft (OHC) pressure cylinder head with removable bearing caps, which of the following is used to measure bearing alignment?

    A. A straightedge and a feeler gauge
    B. Plastigauge
    C. A dial indicator
    D. A telescoping gauge

    Answer A is correct. A straightedge and a feeler gauge can be used to check alignment.

    Answer B is incorrect. Plastigage can be used to measure bearing clearance, but it cannot effectively measure bearing alignment.

    Answer C is incorrect. A dial indicator can be used to measure crankshaft runout, but it cannot measure bearing alignment.

    Answer D is incorrect. A telescoping gauge can be used to measure cylinder taper, but it is not used to measure bearing alignment.

TASK A.9

43. Air is heard escaping from the exhaust during a cylinder leakage test. Technician A says this condition could indicate a leaking exhaust valve. Technician B says the technician must ensure that the cylinder is at top dead center (TDC) of the compression stroke when performing this test. Who is correct?

    A. A only
    B. B only
    C. Both A and B
    D. Neither A nor B

    Answer A is incorrect. Technician B is also correct.

    Answer B is incorrect. Technician A is also correct.

    Answer C is correct. Both Technicians are correct. When performing a cylinder leakage test, the cylinder being tested must be on TDC of the compression stroke. Air heard leaking from the exhaust during a cylinder leakage test is usually an indication of a leaking exhaust valve. If the cylinder is on TDC exhaust instead of TDC compression, the exhaust valve will be open, and air will be heard escaping from the exhaust.

    Answer D is incorrect. Both Technicians are correct.

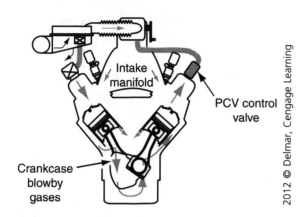

Intake manifold

PCV control valve

Crankcase blowby gases

2012 © Delmar, Cengage Learning

44. The above illustration shows the PCV system. All of the following are symptoms of a stuck open PCV valve EXCEPT:

A. High manifold vacuum

B. Engine stalling

C. Rough idle operation

D. Lean air/fuel ratio

TASK G.2

Answer A is correct. A stuck open valve can cause a low manifold vacuum; it is a vacuum leak.

Answer B is incorrect. Engine stalling may be caused by a PCV valve stuck in the open position, which acts like a vacuum leak.

Answer C is incorrect. Rough idle may occur if a PCV valve is stuck in the open position, because manifold vacuum will be lowered.

Answer D is incorrect. A stuck open valve creates a vacuum leak, admitting unmeasured air and resulting in a lean air/fuel mixture.

45. Which of these should be performed first when a starter fails to crank?

A. Measure battery voltage.

B. Remove spark plugs.

C. Check for the presence of spark.

D. By-pass the starter solenoid with the remote starter button.

TASK A.18

Answer A is correct. Fully charged batteries have an open circuit voltage of 12.68 volts. The technician should start here to verify that there is enough power to turn the starter over.

Answer B is incorrect. Spark plugs might be removed if the engine appears to have a hydro-lock condition due to coolant or fuel in the cylinders.

Answer C is incorrect. Spark has no relevance to cranking.

Answer D is incorrect. This would not be the first check. However, the starter solenoid may be activated by a starter button if the technician suspects that the ignition switch or Neutral safety switch is at fault.

TASK H.7

46. Technician A says that measuring voltage drop can only be performed on a circuit that has current flow. Technician B says a high voltage drop reading indicates excessive resistance in the circuit. Who is correct?

    A. A only

    B. B only

    C. Both A and B

    D. Neither A nor B

Answer A is incorrect. Technician B is also correct.

Answer B is incorrect. Technician A is also correct.

Answer C is correct. Both Technicians are correct. When a voltage drop test is being performed, the circuit must be active, with current flowing through the circuit. A DMM is used with a lead at each end of the wire that is being tested. A high voltage drop reading indicates that voltage is being used up by excessive resistance in the circuit.

Answer D is incorrect. Both Technicians are correct.

TASK D.5

47. A loose belt may cause all of these EXCEPT:

    A. A discharged battery

    B. Water pump bearing failure

    C. Poor power steering assist

    D. Engine overheating

Answer A is incorrect. A loose belt may cause the battery to become discharged due to inadequate alternator output.

Answer B is correct. Water pump bearings will not fail due to a loose belt.

Answer C is incorrect. A loose belt could diminish power steering assist.

Answer D is incorrect. If the belt that drives the water pump is slipping, the water pump may fail to move coolant effectively, which could cause the engine to overheat.

TASK A.3

48. Engine oil is found to be leaking from many different locations on an engine. Technician A says the PCV valve may be restricted. Technician B says the ignition system may be misfiring. Who is correct?

    A. A only

    B. B only

    C. Both A and B

    D. Neither A nor B

Answer A is correct. Only Technician A is correct. A restricted PCV valve can cause increased pressure within the engine crankcase and engine oil leaks from many different locations.

Answer B is incorrect. A misfiring ignition system would not cause engine oil leaks.

Answer C is incorrect. Only Technician A is correct.

Answer D is incorrect. Technician A is correct.

49. When measured on the supply line to the fuel rail, fuel pump flow volume is less than specified. Which of the following could be the cause?

    A. Restricted fuel injectors

    B. An electrically open fuel injector

    C. Restricted return line

    D. Low voltage to the fuel pump

TASK F.6

Answer A is incorrect. Restricted injectors can cause low power, but they will not affect pump volume.

Answer B is incorrect. An electrically open injector can cause a misfire, but it will not affect fuel pump volume.

Answer C is incorrect. A restricted return line will not affect volume if the check is being performed at the supply side of the fuel system.

Answer D is correct. Low voltage to the pump can cause low pump speed and low volume.

50. Technician A says blue-gray smoke coming from the exhaust may be caused by stuck piston rings. Technician B says this could be caused by a plugged oil drain passage in the cylinder head. Who is correct?

    A. A only

    B. B only

    C. Both A and B

    D. Neither A nor B

TASK A.5

Answer A is incorrect. Technician B is also correct.

Answer B is incorrect. Technician A is also correct.

Answer C is correct. Both Technicians are correct. Both stuck piston rings and a plugged oil drain passage in the cylinder head may allow excessive oil to enter the cylinders. This oil would produce blue-gray smoke when it burned.

Answer D is incorrect. Both Technicians are correct.

PREPARATION EXAM ANSWER SHEET FORMS

ANSWER SHEET

1. _____	21. _____	41. _____
2. _____	22. _____	42. _____
3. _____	23. _____	43. _____
4. _____	24. _____	44. _____
5. _____	25. _____	45. _____
6. _____	26. _____	46. _____
7. _____	27. _____	47. _____
8. _____	28. _____	48. _____
9. _____	29. _____	49. _____
10. _____	30. _____	50. _____
11. _____	31. _____	
12. _____	32. _____	
13. _____	33. _____	
14. _____	34. _____	
15. _____	35. _____	
16. _____	36. _____	
17. _____	37. _____	
18. _____	38. _____	
19. _____	39. _____	
20. _____	40. _____	

ANSWER SHEET

1. _____	21. _____	41. _____
2. _____	22. _____	42. _____
3. _____	23. _____	43. _____
4. _____	24. _____	44. _____
5. _____	25. _____	45. _____
6. _____	26. _____	46. _____
7. _____	27. _____	47. _____
8. _____	28. _____	48. _____
9. _____	29. _____	49. _____
10. _____	30. _____	50. _____
11. _____	31. _____	
12. _____	32. _____	
13. _____	33. _____	
14. _____	34. _____	
15. _____	35. _____	
16. _____	36. _____	
17. _____	37. _____	
18. _____	38. _____	
19. _____	39. _____	
20. _____	40. _____	

ANSWER SHEET

1. _____	21. _____	41. _____
2. _____	22. _____	42. _____
3. _____	23. _____	43. _____
4. _____	24. _____	44. _____
5. _____	25. _____	45. _____
6. _____	26. _____	46. _____
7. _____	27. _____	47. _____
8. _____	28. _____	48. _____
9. _____	29. _____	49. _____
10. _____	30. _____	50. _____
11. _____	31. _____	
12. _____	32. _____	
13. _____	33. _____	
14. _____	34. _____	
15. _____	35. _____	
16. _____	36. _____	
17. _____	37. _____	
18. _____	38. _____	
19. _____	39. _____	
20. _____	40. _____	

ANSWER SHEET

1. _____	21. _____	41. _____
2. _____	22. _____	42. _____
3. _____	23. _____	43. _____
4. _____	24. _____	44. _____
5. _____	25. _____	45. _____
6. _____	26. _____	46. _____
7. _____	27. _____	47. _____
8. _____	28. _____	48. _____
9. _____	29. _____	49. _____
10. _____	30. _____	50. _____
11. _____	31. _____	
12. _____	32. _____	
13. _____	33. _____	
14. _____	34. _____	
15. _____	35. _____	
16. _____	36. _____	
17. _____	37. _____	
18. _____	38. _____	
19. _____	39. _____	
20. _____	40. _____	

ANSWER SHEET

1. _____   21. _____   41. _____
2. _____   22. _____   42. _____
3. _____   23. _____   43. _____
4. _____   24. _____   44. _____
5. _____   25. _____   45. _____
6. _____   26. _____   46. _____
7. _____   27. _____   47. _____
8. _____   28. _____   48. _____
9. _____   29. _____   49. _____
10. _____   30. _____   50. _____
11. _____   31. _____
12. _____   32. _____
13. _____   33. _____
14. _____   34. _____
15. _____   35. _____
16. _____   36. _____
17. _____   37. _____
18. _____   38. _____
19. _____   39. _____
20. _____   40. _____

ANSWER SHEET

1. _____
2. _____
3. _____
4. _____
5. _____
6. _____
7. _____
8. _____
9. _____
10. _____
11. _____
12. _____
13. _____
14. _____
15. _____
16. _____
17. _____
18. _____
19. _____
20. _____

21. _____
22. _____
23. _____
24. _____
25. _____
26. _____
27. _____
28. _____
29. _____
30. _____
31. _____
32. _____
33. _____
34. _____
35. _____
36. _____
37. _____
38. _____
39. _____
40. _____

41. _____
42. _____
43. _____
44. _____
45. _____
46. _____
47. _____
48. _____
49. _____
50. _____

Glossary

Absolute Pressure The zero point from which pressure is measured.

Actuator A device that delivers motion in response to an electrical or hydraulic signal.

Accelerator Pedal Position (APP) Sensor The sensor that sends the position of the accelerator pedal to the engine control module (ECM).

A/D Converter Acronym for analog-to-digital converter.

Additive An ingredient intended to improve a certain characteristic of a material or fluid.

After-Cooler Any of a number of types of charge air cooler.

Air Compressor An engine-driven mechanism for supplying compressed air to the truck brake system.

Air Conditioning The control of air movement, humidity, and temperature by mechanical means.

Air Dryer A unit that removes moisture.

Air Filter A device that minimizes the possibility of impurities entering the induction system.

Altitude Compensation System An altitude barometric switch and solenoid used to provide better driveability at altitudes exceeding 1,000 feet.

Ambient Temperature Temperature of the surrounding or prevailing air. Normally, it is considered to be the temperature in the service area where testing is taking place.

Amp Abbreviation for ampere.

Ampere The unit for measuring electrical current.

Analog Signal A voltage signal that varies within a given range (from high to low, including all points in between).

Analog-to-Digital Converter (A/D converter) A device that converts analog voltage signals to a digital format; this is located in a section of the processor called the input signal conditioner.

Analog Volt/Ohmmeter (AVOM) A test meter used for checking voltage and resistance. Analog meters should not be used on solid state circuits.

Antifreeze A compound, such as alcohol or glycerin, added to water to lower its freezing point.

Antirust Agent Additive used with lubricating oils to prevent rusting of metal components when the engine is not in use.

Armature The rotating component of a (1) starter or other motor, (2) generator, (3) compressor clutch.

ASE Automotive Service Excellence, a trademark of National Institute for Automotive Service Excellence.

ATDC After top dead center.

Atmospheric Pressure The weight of the air at sea level; 14.696 pounds per square inch (psi) or 101.33 kilopascals (kPa).

Auxiliary Filter Any type of supplementary filter.

Axis of Rotation The center line around which a gear or part revolves.

Backpressure The amount of restriction present in the exhaust system. Some backpressure is normal, but excess backpressure reduces the efficiency of the engine.

Battery Terminal A tapered post or threaded studs on top of the battery case, or infernally threaded provisions on the side of the battery for connecting the cables.

Bellows A movable cover or seal that is pleated or folded like an accordion to allow for expansion and contraction.

Bimetallic Two dissimilar metals joined together that have different bending characteristics when subjected to changes of temperature.

Blade Fuse A type of fuse having two flat male lugs sticking out for insertion in the female box connectors.

Block Diagnosis Chart A troubleshooting chart that lists symptoms, possible causes, and probable remedies in columns.

Blowby The combustion gases that leak past the piston rings and enter the crankcase. Most blowby is contained by the PCV system, so excessive blowby is an indication of worn rings or cylinders.

Blower Fan A fan that pushes or blows air through a ventilation, a heater, or an air-conditioning system.

Boss A race of a bearing or shaft bore.

Bottom Dead Center (BDC) The piston is at its lowest point in the cylinder.

Bottoming A condition that occurs when the teeth of one gear touch the lowest point between teeth of a mating gear.

Bracket An attachment used to secure components to the body or frame.

BTDC Before top dead center.

British thermal unit (Btu) A measure of heat quantity equal to the amount of heat required to raise 1 pound of water 1°F.

Btu British thermal unit.

Camshaft Position (CMP) Sensor Used to send the position of the camshaft to the ECM.

Cartridge Fuse A type of fuse having a strip of low melting point metal enclosed in a glass tube. If an excessive current flows through the circuit, the fuse element melts at the narrow portion opening the circuit and preventing damage.

Cavitation A condition caused by vapor bubble collapse.

Ceramic Fuse A fuse found in some import vehicles that has a ceramic insulator with a conductive metal strip along one side.

Charging Circuit The alternator (or generator) and associated circuit used to keep the battery charged and power the vehicle electrical systems when the engine is running.

Charging System A system consisting of the battery, alternator, voltage regulator, associated wiring, and the electrical loads of a vehicle. The purpose of the system is to recharge the battery whenever necessary and to provide the current required to power the electrical components.

Check-Valve A valve that allows airflow in one direction only.

Circuit The complete path of an electrical current, including the generating device. When the path is unbroken, the circuit is closed and current flows. When the circuit continuity is broken, the circuit is open and current flow stops.

Closed Loop The mode of operation when the oxygen sensor is sending a signal to the ECM indicating the quantity of oxygen in the exhaust stream, and the ECM is correcting fuel delivery based upon this measurement.

Crankshaft Position (CKP) Sensor Used to signal the position of the crankshaft to the ECM.

COE Cab-over-engine.

Coefficient of Friction A measurement of friction developed between two objects in contact when one of the objects is drawn across the other.

Combination A truck coupled to one or more trailers.

Compression Applying pressure to a spring or any springy substance, thus causing it to reduce its length in the direction of the compressing force.

Compressor (1) A mechanical device that increases pressure within a container by pumping air into it. (2) That component of an air conditioning system that compresses low temperature/pressure refrigerant vapor.

Condensation The process by which gas (or vapor) changes to a liquid.

Condenser A component in an air conditioning system used to cool a refrigerant below its boiling point, causing it to change from a vapor to a liquid.

Conductor Any material that permits electrical current flow.

Coolant Liquid that circulates in an engine cooling system.

Coolant Heater A component used to aid engine starting and reduce wear caused by cold starting.

Coolant Hydrometer A tester designed to measure coolant specific gravity and determine the amount of antifreeze protection.

Cooling System Complete system for circulating coolant.

COP Coil-on-plug ignition system.

Crankcase Housing within which the crankshaft and other components of the engine operate.

Crankcase Pressure Crankcase pressure is the result of combustion pressure that escapes past the rings. Crankcase pressure should be removed by a properly functioning PCV system and there should be a slight vacuum.

Cranking Circuit The starter and its associated circuit, including battery, relay (solenoid), ignition switch, neutral start switch (on vehicles with automatic transmission), cables, and wires.

Cycling (1) Repeated on-off action of an air or A/C compressor. (2) Repeated full electrical discharge and recharge of a battery that can cause the positive plate material to break away from its grids and fall into the sediment chambers at the base of the battery case.

Dampen To slow or reduce oscillations or movement.

Dash Control Valves A variety of hand-operated valves located on the dash. They include parking brake valves, tractor protection valves, and differential lock.

Data Links Circuits through which computers communicate with other electronic devices such as control panels, modules, some sensors, or other computers in the form of digital signals.

Data Link Connector (**DLC**) The connector on the vehicle that the technician uses to connect the scan tool to the data bus.

Deadline To take a vehicle out of service.

Deburring To remove sharp edges from a cut.

Deflection Bending or moving to a new position as the result of an external force.

Department of Transportation (DOT) Government agency that establishes vehicle standards.

Detergent Additive Oil additive that helps keep metal surfaces clean and prevents deposits. These additives suspend particles of carbon and oxidized oil in the oil.

Detonation A destructive condition in the cylinder that occurs when there are multiple flame fronts that collide, producing a metallic knock during combustion. The initial flame front is produced by the firing of the spark plug, and secondary flame fronts are the result of either spontaneous or compression ignition. Detonation causes high cylinder pressure spikes that can cause serious damage to the engine.

DER Department of Environmental Resources.

Diagnostic Flow Chart A chart that provides a systematic approach to electrical system and component troubleshooting and repair. Found in service manuals and are vehicle make and model specific.

Dial Caliper A measuring instrument capable of taking inside, outside, depth, and step measurements.

Digital Binary Signal A signal that has only two values: on and off.

Digital Multi-Meter (DMM) A single tool which can be used to measure volts, ohms, and amps.

Digital Volt/Ohmmeter (DVOM) A type of test meter recommended by most manufacturers for use on solid state circuits.

Diode The simplest semiconductor device formed by joining P-type semiconductor material with N-type semiconductor material. A diode allows current to flow in one direction but not in the opposite direction.

DIS Distributorless ignition system.

Dispatch Sheet Form used to keep track of dates when work is to be completed. Some dispatch sheets follow the job through each step of the servicing process.

DOT Department of Transportation.

Drive or Driving Gear A gear that drives another gear or causes another gear to turn.

Driveline The propeller or drive shaft and universal joints that link the transmission to the axle pinion gear shaft.

Driven Gear A gear that is driven or forced to turn by a drive gear, by a shaft, or by some other device.

DTC Diagnostic trouble code.

ECM Engine control module.

Engine Coolant Temperature (ECT) Sensor Normally a non-linear negative temperature coefficient thermistor, which has high resistance when cold and low resistance when hot.

ECU Electronic control unit.

EGR Exhaust gas recirculation valve.

Electricity Movement of electrons from one place to another.

Electric Retarder Electromagnets mounted in a steel frame. Energizing the retarder causes the electromagnets to exert a dragging force on the rotors in the frame, and this drag force is transmitted directly to the drive shaft.

Electromotive Force (EMF) The force that moves electrons between atoms. This force is the pressure that exists between the positive and negative points (the electrical imbalance). This force is measured in units called volts.

Electronic Control Unit (ECU) An electronic system controller.

Electronics The technology of moving electrons through hard wire, semiconductor, gas, and vacuum circuits.

Electrons Negatively charged particles orbiting around every atomic nucleus.

EMF Electromotive force.

Engine Brake A hydraulically operated device that converts the vehicle engine into a power-absorbing retarding mechanism.

Engine Stall Point The point, in RPMs, underload is compared to the engine manufacturer's specified RPM for the stall test.

Engine Off Natural Vacuum (EONV) System Uses the natural vacuum created in the fuel supply system to control evaporative emissions.

Environmental Protection Agency United States government agency charged with the responsibilities of protecting the environment and enforcing the Clean Air Act (CAA) of 1990.

EPA Environmental Protection Agency.

EPROM Electronically programmable memory.

EVAP System The evaporative control system. This system is used to control the evaporative emissions of gasoline.

Exhaust Brake A valve in the exhaust pipe between the manifold and the muffler. A slide mechanism that restricts the exhaust flow, causing exhaust backpressure to build up in the engine's cylinders. The exhaust brake actually transforms the engine into a low-pressure air compressor driven by the wheels.

False Brinelling Polishing of a surface that is not damaged.

Fatigue Failures Progressive destruction of a shaft or gear teeth material usually caused by overloading.

Fault Code A code recorded to computer memory. A fault code can be read by a digital diagnostic reader. *See Diagnostic Trouble Code (DTC).*

Feeler Gauge A tool containing strips of metal in varying thickness, typically ranging from 0.0015" to 0.025".

Fixed Value Resistor An electrical device designed to have only one resistance rating, which should not change, for controlling voltage.

Flammable Any material that will easily catch fire or explode.

Flare To spread gradually outward in a bell shape.

Flex Disc A term often used for flexplate.

Flexplate Component used to mount the torque converter to the crankshaft. The purpose of the flexplate is to transfer crankshaft rotation to the torque converter assembly.

Float A cruising drive mode in which throttle setting matches engine speed to road speed, neither accelerating nor decelerating.

Foot-Pound An English unit of measurement for torque. One foot-pound is the torque obtained by a force of one pound applied to a foot-long wrench handle.

Franchised Dealership A dealer that has signed a contract with a particular manufacturer to sell and service a particular line of vehicles.

Fretting A result of vibration that the bearing outer race can pick up the machining pattern.

Fuse Link A short length of smaller gauge wire installed in a conductor, usually close to the power source.

Fusible Link A term often used for fuse link.

GCW Gross combination weight.

Gear A disc-like wheel with external or internal teeth that serves to transmit or change motion.

Gear Pitch The number of teeth per given unit of pitch diameter; an important factor in gear design and operation.

Gross Combination Weight (GCW) The total weight of a fully quipped vehicle including payload, fuel, and driver.

Gross Trailer Weight (GTW) The sum of the weight of an empty trailer and its payload.

Gross Vehicle Weight (GVW) The total weight of a fully equipped vehicle and its payload.

Ground The negatively charged side of a circuit. A ground can be a wire, the negative side of the battery, or the vehicle chassis.

Grounded Circuit A shorted circuit that causes current to return to the battery before it has reached its intended destination.

GTW Gross trailer weight.

GVW Gross vehicle weight.

Hall Effect A sensor which generates a digital signal. Typically, this type of sensor can be found on some engine position sensors.

Harness and Harness Connectors The organization of the vehicle's electrical system, providing an orderly and convenient starting point for tracking and testing circuits.

Hazardous Materials Any substance that is flammable, explosive, or is known to produce adverse health effects in people or the environment that are exposed to the material during its use.

Heads-Up Display (HUD) A technology used in some vehicles that superimposes data on the driver's normal field of vision. The operator can view the information, which appears to "float" just above the hood at a range near the front of a conventional tractor or truck. This allows the driver to monitor conditions such as limited road speed without interrupting his normal view of traffic.

Heater Control Valve A valve that controls the flow of coolant into the heater core from the engine.

Heat Exchanger A device used to transfer heat, such as a radiator or condenser.

Heavy-Duty Truck A truck that has a GVW of 26,001 pounds or more.

High-Resistant Circuits Circuits that have an increase in circuit resistance, with a corresponding decrease in current.

High-Strength Steel A middle-alloy steel that is much stronger than hot-rolled or cold-rolled sheet steels that normally are used in the manufacture of car body frames.

Hinged Pawl Switch The simplest type of switch; one that makes or breaks the current of a single conductor.

HO₂S Heated oxygen sensor mounted in the exhaust stream to calculate the oxygen content.

HUD Heads-up display.

Hydrometer A tester designed to measure the specific gravity of a liquid.

Inboard Toward the center line of the vehicle.

In-Line Fuse A fuse in series with the circuit in a small plastic fuse holder, not in the fuse box or panel. Used as a protection device for a portion of the circuit even though the entire circuit may be protected by a fuse in the fuse box or panel.

Installation Templates Drawings supplied by some vehicle manufacturers to allow the technician to install the accessory correctly. The templates available can be used to check clearances or to ease installation.

Insulator A material, such as rubber or glass, that offers high resistance to the flow of electrons.

Integrated Circuit A component containing diodes, transistors, resistors, capacitors, and other electronic components mounted on a single piece of material and capable of performing numerous functions.

Jumper Wire A wire used to temporarily by-pass a circuit or components for electrical testing. A jumper wire consists of a length of wire with an alligator clip at each end.

Jump Out A condition that occurs when a fully engaged gear and sliding clutch are forced out of engagement.

Jump Start Procedure used when it becomes necessary to use a booster battery to start a vehicle having a discharged battery.

Kinetic Energy Energy in motion.

KOEO Acronym for key on, engine off.

Lateral Runout The wobble or side-to-side movement of a rotating wheel or of a rotating wheel and tire assembly.

Less Than Truckload (LTL) Partial loads from the networks of consolidation centers and satellite terminals.

Linkage A system of rods and levers used to transmit motion or force.

Low-Maintenance Battery A conventionally vented, lead/acid battery, requiring normal periodic maintenance.

LTL Less than truckload.

Magnetorque An electromagnetic clutch.

Maintenance-Free Battery A battery that does not require the addition of water during normal service life.

Maintenance Manual A publication containing routine maintenance procedures and intervals for vehicle components and systems.

Manifold Absolute Pressure (MAP) Sensor Sends the intake manifold vacuum reading to the ECM, which uses this value to help calculate engine load.

Margin Area between the valve face and the head of the valve.

MIL Malfunction indicator lamp; also known as the check engine lamp.

NATEF National Automotive Technicians Education Foundation.

National Automotive Technicians Education Foundation (NATEF) Foundation having a program of certifying secondary and postsecondary automotive and heavy-duty truck training programs.

National Institute for Automotive Service Excellence (ASE) A nonprofit organization that has an established certification program for automotive, heavy-duty truck, auto body repair, engine machine shop technicians, and parts specialists.

Needlenose Pliers This tool has long tapered jaws for grasping small parts or for reaching into tight spots. Many needlenose pliers also have cutting edges and a wire stripper.

NIASE National Institute for Automotive Service Excellence, now abbreviated ASE.

NIOSH National Institute for Occupation Safety and Health.

NLGI National Lubricating Grease Institute.

NHTSA National Highway Traffic Safety Administration.

OBD-II On-board diagnostic system 2.

OEM Original equipment manufacturer.

Off-road With reference to unpaved, rough, or ungraded terrain on which a vehicle will operate. Any terrain not considered part of the highway system falls into this category.

OHC Overhead camshaft.

Ohm A unit of measured electrical resistance.

Ohm's Law The basic law of electricity stating that in any electrical circuit, current, resistance, and pressure work together in a mathematical relationship.

OHV Overhead valve.

On-road With reference to paved or smooth-graded surface terrain on which a vehicle will operate, generally considered to be part of the public highway system.

Open Circuit An electrical circuit whose path has been interrupted or broken either accidentally (a broken wire) or intentionally (a switch turned off).

Open Loop The mode of operation when the oxygen sensor is not sending a valid signal to the ECM, and the ECM is controlling fuel delivery based primarily on temperature, engine load, and accelerator pedal position.

Oscillation The rotational movement in either fore/aft or side-to-side direction about a pivot point.

Oscilloscope A tool used to display voltage across a time scale.

OSHA Occupational Safety and Health Administration.

Oval A condition that occurs when a tube or bore is not round, but somewhat egg-shaped.

Overspeed Governor A governor that shuts off the fuel or stops the engine when excessive speed is reached.

Oxidation Inhibitor (1) An additive used with lubricating oils to keep oil from oxidizing even at very high temperatures. (2) An additive for gasoline to reduce the chemicals in gasoline that react with oxygen.

Parallel Circuit An electrical circuit that provides two or more paths for current flow.

Parts Requisition Form used to order parts, on which the technician writes the names of what part(s) are needed along with the vehicle's VIN or company's identification folder.

Payload Weight of the cargo carried by a truck, not including the weight of the body.

PCM Powertrain control module.

Positive Crankcase Ventilation (PCV) System Removes the crankcase vapors from the engine crankcase, allowing them to enter the intake manifold.

PCV Valve Valve which controls the airflow of the PCV system.

Pitting Surface irregularities resulting from corrosion.

Plastigauge® A tool used to measure clearances. Typically used to measure engine bearing clearances.

Polarity The particular state, either positive or negative, with reference to the two poles or to electrification.

Pole The number of input circuits made by an electrical switch.

Pounds per Square Inch (psi) A unit of English measure for pressure.

Power A measure of work being done factored with time.

Power Balance Test A test in which either fuel or spark is removed from a cylinder and the technician watches the resulting impact on engine operation.

Pressure The amount of force applied to a definite area measured in pounds per square inch (psi) English or kilopascals (kPa) metric.

Pressure Differential Difference in pressure between two points of a system or a component.

Printed Circuit Board An electronic circuit board made of nonconductive material onto which conductive metal, such as copper, has been deposited. Parts of the metal are then etched away by an acid, leaving metal lines that form the conductors for the various circuits on the board. A printed circuit board can hold many complex circuits in a very small area.

Priority Valve A valve that ensures that the control system upstream from the valve will have sufficient pressure during shifts to perform its automatic functions.

Programmable Read-Only Memory (PROM) An electronic component that contains program information specific to different vehicle model calibrations.

PROM Programmable read-only memory.

psi Pounds per square inch.

Pull Circuit A circuit that brings the cab from a fully tilted position up and over the center.

Pulse Width (PW) The measurement of an electric circuit on time. Injector on time is typically referred to as pulse width.

Push Circuit A circuit that raises the cab from the lowered position to the desired tilt position.

P-type Semiconductors Positively charged materials, which enables them to carry current. They are produced by adding an impurity with three electrons in the outer ring (trivalent atoms).

Radial Load A load that is applied at 90 degrees to an axis of rotation.

RAM Random-access memory.

Ram Air Air that is forced into the engine or passenger compartment by the forward motion of the vehicle.

Random-Access Memory (RAM) The memory used during computer operation to store temporary information. The microcomputer can write, read, and erase information from RAM in any order, which is why it is called random.

Rated Capacity The maximum, recommended safe load that can be sustained by a component or an assembly without permanent damage.

RCRA Resource Conservation and Recovery Act.

Reactivity The characteristic of a material that enables it to react with air, heat, water, or other materials.

Read-Only Memory (ROM) A type of memory used in microcomputers to store information permanently.

Recall Bulletin A bulletin that pertains to special situations that involve service work or replacement of components in connection with a recall notice.

Reference Voltage The voltage, supplied to a sensor by the computer, that acts as a baseline voltage; modified by the sensor to act as an input signal.

Refractometer A tool used to measure the specific gravity of a liquid.

Relay An electric switch that allows a small current to control a much larger one. It consists of a control circuit and a power circuit.

Reserve Capacity Rating The ability of a battery to sustain a minimum vehicle electrical load in the event of a charging system failure.

Resistance The opposition to current flow in an electrical circuit.

Revolutions per Minute (RPM) The number of complete turns a member makes in one minute.

Right to Know Law A law passed by the federal government and administered by the Occupational Safety and Health Administration (OSHA) that requires any company that uses or produces hazardous chemicals or substances to inform its employees, customers, and vendors of any potential hazards that may exist in the workplace as a result of using the products.

ROM Read-only memory.

Rotation A term used to describe the fact that a gear, shaft, or other device is turning.

RPM Revolutions per minute.

Rotor (1) A part of the alternator that provides the magnetic fields necessary to create a current flow. (2) The rotating member of an assembly.

Runout Deviation or wobble in a shaft or wheel as it rotates. Runout is measured with a dial indicator.

Screw Pitch Gauge A gauge used to provide a quick and accurate method of checking the threads per inch of a nut or bolt.

Semiconductor A solid state device that can function as either a conductor or an insulator, depending on how its structure is arranged.

Semitrailer A load-carrying vehicle equipped with one or more axles and constructed so that the trailer's front end is supported on the fifth wheel of the truck tractor that pulls it.

Sensing Voltage The voltage that allows the regulator to sense and monitor the battery voltage level.

Sensor An electronic device used to monitor conditions for computer control requirements.

Series Circuit A circuit that consists of two or more resistors connected to a voltage source with only one path for the electrons to follow.

Series/Parallel Circuit A circuit designed so that both series and parallel combinations exist within the same circuit.

Service Bulletin A publication that provides the latest service tips, field repairs, product improvements, and related information of benefit to service personnel.

Service Manual A manual, published by the manufacturer, that contains service and repair information for vehicle systems and components.

Short Circuit An undesirable connection between two worn or damaged wires.

Solenoid An electromagnet used to perform mechanical work, made with one or two coil windings wound around an iron tube.

Solid Wire A single-strand conductor.

Solvent A substance that dissolves other substances.

Spade Fuse A term used for blade fuse.

Spalling Surface fatigue occurs when chips, scales, or flakes of metal break off due to fatigue rather than wear.

Specialty Service Shop A shop that specializes in areas such as engine rebuilding, transmission/axle overhauling, brake, air conditioning/heating repairs, and electrical/electronic work.

Specific Gravity The measurement of a liquid based on the ratio of the liquid's mass to an equal volume of distilled water.

Spontaneous Combustion A process by which a combustible material ignites by itself.

Stall Test A test performed when there is a malfunction in the vehicle power package (engine and transmission), to determine which of the components is at fault.

Starter Circuit The circuit that carries the high current flow within the system and supplies power for the actual engine cranking.

Starter Motor Device that converts electrical energy from the battery into mechanical energy for cranking the engine.

Starter Safety Switch A switch that prevents vehicles with automatic transmissions from being started in gear.

Static Balance Balance at rest, or still balance.

Stepped Resistor A resistor designed to have two or more fixed values, available by connecting wires to one of several taps.

Storage Battery Source of direct current electricity for both the electrical and electronic systems.

Straightedge A piece of flat stock that has been precision ground to be flat. This tool is used in conjunction with a feeler gauge to determine if a component is warped or otherwise not flat.

Stranded Wire Wire that is made up of a number of small solid wires, generally twisted together to form a single conductor.

Sulfation A condition that occurs when sulfate is allowed to remain in the battery plates for a long time, causing two problems. (1) It lowers the specific gravity levels, increasing the danger of freezing at low temperatures. (2) In cold weather a sulfated battery may not have the reserve power needed to crank the engine.

Swage To reduce or taper.

Switch Device used to control on/off and direct the flow of current in a circuit. A switch can be under the control of the driver or can be self-operating through a condition of the circuit, the vehicle, or the environment.

Tachometer An instrument that indicates rotating speeds, sometimes used to indicate crankshaft RPM.

Tapped Resistor A resistor designed to have two or more fixed values, available by connecting wires to one of several taps.

Technical Service Bulletin (TSB) Bulletin issued by the vehicle manufacturer to alert service technicians of revised service procedures.

Throttle Position Sensor (TPS or TP) A sensor that sends the position of the throttle plate to the ECM.

Throw (1) The offset of a crankshaft. (2) The number of output circuits of a switch.

Time Guide Prepared reference material used for computing compensation payable by the truck manufacturer for repairs or service work to vehicles under warranty, or for other special conditions authorized by the company.

Timing (1) Procedure of marking the appropriate teeth of a gear set prior to installation and placing them in proper mesh while in the transmission. (2) Combustion spark delivery in relation to the piston position.

Top Dead Center (TDC) The piston is at its highest point in the cylinder.

Torque Twisting force to tighten a fastener to a specific degree of tightness, generally in a given order or pattern.

Toxicity A statement of how poisonous a substance is.

Tractor A motor vehicle, without a body, that has a fifth wheel and is used for pulling a semitrailer.

Transistor Three terminal semiconductors used for switching and amplifying electronic circuits.

Tree Diagnosis Chart A chart used to provide a sequence for what should be inspected or tested when troubleshooting a repair problem.

TTY Acronym for torque to yield. Generally refers to head bolts, which are torqued to just below their yield point, and should not be reused.

Vacuum Absence of matter, but often used to describe any pressure condition below atmospheric pressure.

Validity List A list supplied by the manufacturer of valid bulletins.

Valve Spring Installed Height The distance from the bottom of the valve spring retainer and the spring seat on the cylinder head. It should be checked while reconditioning the cylinder head since a valve spring installed height in excess of specifications lengthens the effective length of the spring, reducing pressure.

VIN Vehicle identification number.

Volt The unit of electromotive force.

Voltage-Generating Sensors Devices that produce their own voltage signal.

Voltage Limiter A device that provides protection by limiting voltage to the instrument panel gauges to approximately 5 volts.

Voltage Regulator A device that controls the amount of current produced by the alternator or generator and thus the voltage level in the charging circuit.

Watt The measure of electrical power.

Watt's Law A basic law of electricity used to find the power of an electrical circuit expressed in watts. It states that power equals the voltage multiplied by the current, in amperes.

Windings (1) The three separate bundles in which wires are grouped in the stator. (2) The coil of wire found in a relay or other similar device. (3) That part of an electrical clutch that provides a magnetic field.

Work (1) Forcing a current through a resistance. (2) The product of a force.

Yield Strength The highest stress a material can stand without permanent deformation or damage, expressed in pounds per square inch (psi).

Zener Diode A variation of the diode, this device functions like a standard diode until a certain voltage is reached. When the voltage level reaches this point, the zener diode will allow current to flow in the reverse direction. Zener diodes are often used in electronic voltage regulators.

Notes

Notes

Notes

Notes

Notes

Notes